Emergy

I dedicate this book:

to my sister – may her health spare her a little;

to her husband and children, may they be
philosophical in the face of life's challenges;

to all of my own lecturers

"The wise man does not aspire to pleasure,
but to the absence of suffering."

Aristotle

Thermodynamics – Energy, Environment, Economy Set
coordinated by
Michel Feidt

Emergy

Olivier Le Corre

First published 2016 in Great Britain and the United States by ISTE Press Ltd and Elsevier Ltd

ISTE Press Ltd
27-37 St George's Road
London SW19 4EU
UK

www.iste.co.uk

Elsevier Ltd
The Boulevard, Langford Lane
Kidlington, Oxford, OX5 1GB
UK

www.elsevier.com

Notices

Knowledge and best practice in this field are constantly changing. As new research and experience broaden our understanding, changes in research methods, professional practices, or medical treatment may become necessary.

Practitioners and researchers must always rely on their own experience and knowledge in evaluating and using any information, methods, compounds, or experiments described herein. In using such information or methods they should be mindful of their own safety and the safety of others, including parties for whom they have a professional responsibility.

To the fullest extent of the law, neither the Publisher nor the authors, contributors, or editors, assume any liability for any injury and/or damage to persons or property as a matter of products liability, negligence or otherwise, or from any use or operation of any methods, products, instructions, or ideas contained in the material herein.

For information on all our publications visit our website at http://store.elsevier.com/

British Library Cataloguing-in-Publication Data
A CIP record for this book is available from the British Library
Library of Congress Cataloging in Publication Data
A catalog record for this book is available from the Library of Congress
ISBN 978-1-78548-097-3

Printed and bound in the UK and US

Contents

Acknowledgments

This work would not have been possible without the support and, above all, the friendship, of L. Truffet, Associate Professor at the Ecole des Mines de Nantes[1]. His positive scientific contributions in the emergy sphere, through deploying the tools of formal languages and applied mathematics and technology, remain exceedingly prolific.

The supervision of PhD students or interns requires both accuracy and concision. The author is therefore indebted for all discussions, leading to shared knowledge rather than the ownership of knowledge by given individuals. In particular, the author would like to express his gratitude to N. Jamali-Zghal.

Finally, the author expresses his deep appreciation to M. Feidt. As the instigator of the COFRET conference, his work, scientific thoroughness, openness and scientific curiosity remain exemplary.

---

[1] A French engineering school.

Preface

This work is intended for economic and academic actors who are conscious of implementing or disseminating technical solutions which fall under the scope of a particular development that demonstrates consideration for our planet's limited resources. H.T. Odum laid down the basis of a concept (as much theoretical as practical), that rests upon using the Earth's resources (as much energy as mineral-based). For example, hydrocarbons are the result of the decomposition of lush vegetation in geological eras such as the Cretaceous period, and the supply of heat in the depths of the lithosphere over millions of years. Within a traditional energy approach these hydrocarbons are characterized by their specific heating value and this property is then used to calculate the energy efficiency of particular equipment. For example, the efficiency of a diesel engine may reach 42% using a traditional approach. Likewise, the output of a photovoltaic cell is in the range of 15–20% of direct radiation. Yet this output calculation does not take account of the upstream energy which was used to create hydrocarbons. For H.T. Odum, the energy of hydrocarbons should be changed into a primary energy source (coming, in large part, from the Sun). The output of a diesel engine is thus approximately divided by 2.00E+05, while no adjustment should be made to that of the photovoltaic cell. Moreover, goods and services may also be integrated into such an analysis using conversion factors.

H.T. Odum's theory may be understood as an extension of carbon audits which are already designated by a carbon footprint, or as having certain similarities to lifecycle analysis. By studying the history of resources used within a system or a process, the energy footprint of the aforementioned system is analyzed, giving us the neologism "eMergy".

Emergetic analysis makes stark comparisons between the various energy resources, and in this sense the operation is both a promising and relevant tool, particularly after the COP21 Paris climate talks. Numerous researchers from emerging countries such as China, Brazil and New Zealand are developing strategies on different scales (from single production units to regional or even at nationwide level). The University of Florida, supported by the US Department of Energy, is highly active in this sphere. Researchers in Europe (including Italy, Luxembourg, France and other countries) contribute as much to theoretical concepts as to actual applications.

This work sets out the paradigms of emergy and offers examples for its application. This approach is innovative and improvements and details are constantly broached in works within the field. After about 10 years of research, the author has chosen to start from the historic make-up of emergy and related developments.

The author is a member of the International Society for Advancement of Emergy Research (ISAER). As a Doctor at the Ecole des Mines de Paris[1], he has co-authored 12 first-tier publications and supervised two thesis works within the field.

---

1 One of the top engineering schools in Paris.

Nomenclature

a	Adjuvant
ash	Ashes
C	Capacity of a lorry
Cs	Specific diesel consumption per 100 km
CO_2	Quantity of carbon dioxide
D	Distance
DEF	Domestic emission factor
e	Employees
em	Emergy unit (per product unit)
Em	Total emergy (seJ)
Ex	Total exergy (J)
EF	Exploitation factor
EIR	Emergy investment ratio
ELR	Emergy environmental load ratio
ESI	Emergy sustainability index
EYR	Emergy output ratio
g	Gibb's free energy
Geo	Geo-biosphere

GDP	Gross Domestic Product
H	Humidity
HF	Hubbert's function
i, j, k	"Mute" indices
inv	Investment
I	Maximum number of cycles within the reservoir
K	Gain
l	Length of a blade
LHV	Low heating value
LR	Landfill ratio
m	Mass (g)
M	Molar mass (g/mol)
MFI	Material formation indicator
MPT	Minimum profitability threshold
n	Number of cycles
Nb	Number
NPP	Net primary production
p	Recycling level loss
P	Number of products created
Pg	Geological period
Pu	Purity
PF	Process factor (the so-called "geo-biosphere")
q	Recycled mass fraction
Q	Energy
QR	Quality ratio (recycling)
R	Universal ideal gas constant
RBI	Recycling benefit index
RI	Recyclability indicator

RII	Recycling Interest Index
RYB	Recycle yield ratio (benefit)
s	Form factor
S	Emergy source
SB	Surface swept by the blades
t	Time
T	Conversion
T°	Baseline temperature
Tr	Transformity
TR	Transport (emission factor)
UEV	Unit emergy value
UFEF	Upstream fuel emissions factor
V	Velocity
VR	Velocity ratio
w	Wood
W	Power
x	Energy or mass fraction
y	Molar fraction
z	Height
%m	Mass fraction of a mineral within the earth's crust
%O	Ocean floor
%R	Renewable materials product content

Index

a	Adjuvant
air	Air
c	Recycling process
carb	Carbonification

cc	Concentration
ch	Chemical (bonding)
cm	Recycled material
co	Composition
cv	Conversion
Cr	Earth's crust
e	Eolian
ec	Earth's core
el	Electricity
E	Element
fo	Formulation
fu	Fusion
F	Input (flows) of economic system
G	Imported goods
I	Financial import flows
Inv	Annual investment costs
LF	Landfill facility
m	Mineral
max	Maximum
M	Moon
N	Non-renewable
O	Ocean
p	Output
peak	Peak
pr	Preparation (s+s&s+fu)
r	Refining
rad	Radioactivity
ref	Point of reference

rm	Raw material
rp	Recycled product
R	Renewable
s	Separation (and collection)
s&s	Shredding and sorting
S	Sunlight
th	Thermal
T	Tide
w	Wind

Greek letters

α	Connecting activity
ε	Rate of diesel oxidation
γ	No-load consumption factor
η	Output
ρ	Density
τ	Wood ash content
Δ	Difference
Ψ	Adjustment factor

Mathematical notation

$\langle h \rangle$	Mean of h

Exponents

Cv	Conversion (source supply)
d	Diesel
eco	Eco-conception

f	Fuel
H	Hellman coefficient (0.28)
$H2020$	Reference to a 20% reduction objective
i	Input
LF	Landfill facility
max	Maximum
min	Minimum
ng	Natural gas
ngi	Natural gas transport infrastructure
ngs	Natural gas heating system
o	Output
rc	Release cycle
s	Storage
tr	Transport
w	Wood
wd	Work done
ws	Wood-fired heating system
€	Currency

Introduction

Unfortunately, every year there are:

– 36 billion of tons of carbon dioxide emitted;

– between 130,000 and 150,000 km^2 of forest destroyed;

– between 40 and 250 species disappearing;

– around 65,000 km^2 more desertification;

– 30 billion of oil barrels consumed;

– many other climatic and geo-biospherical occurrences.

These environmental effects are very difficult to quantify in a market economy which is forever endeavoring to grow, indeed for an additional 85 million humans.

"Growth for the sake of growth is the ideology of the cancer cell."

Edward Abbey

The concept of the ecological footprint, that is the number of planets necessary to maintain the living standards of a given population, designates the pressure exerted upon the planet by human activity. For example, in 2014, it "required" the equivalent of four Earths to create a global population with an average standard of living of the inhabitants of the USA or 1.2 with that of an inhabitant of China, or 1.5 Earths to achieve the average global standard. The anthropic impact exceeds the limits of the planet [VAL 15]: exhausting resources, including all resources (whether fossil and mineral), as well as the destruction of the biodiversity requires a significant and unprecedented awareness. A concept known as sustainable development appeared; an oxymoron for some, with multiple definitions for others. The definition proposed in the Bruntland report [WCE 87], written for the UN, has the greatest consensus:

> "*Sustainable development* is development that satisfies current needs without comprising the capacity of future generations to satisfy theirs."

Flowing from this definition three consequences are generally put forward: energy economies, recycling and *renewable energies*. An energy source is commonly labelled as renewable if it remains available on a human scale having, therefore, an ability to work and/or to be consumed, which is less than or equal to its availability. Energy conversion using solar, tidal, wind or geothermic resources is classified as *renewable*. Nuclear energy is classified separately from some features (in particular because of the issue of waste). Fuel wood is not *de facto* a renewable energy source. Thus, the biomass, which has lain for millions of years at the heart of the lithosphere, under intense pressure, is not considered as "renewable", but is, however, the source of fossil fuels. Nevertheless, fuel wood under specific developement conditions is classified as a renewable resource. The quality of low greenhouse gas emissions (of which carbon dioxide is one) is generally associated with renewable resources.

Sustainable development has only one commonly accepted definition, with various methods of evaluation quantifying the environmental impact of an existing technical solution. Such solutions are increasingly divided into two major categories: "product-driven" and "resource-driven" (see Figure I.1). It is not a question of placing these two major families in opposition but of understanding them as complementary, particularly when considering *sustainable development* as not being a shared notion:

– the most widely known "product-driven" method is lifecycle analysis (LCA) [KHA 02]. This multi-criteria method may therefore lead to conclusions according to the weightings opted for;

– professor H.T. Odum introduced the concept of *embodied energy*, leading to the neologism "emergy", so as to designate the energy content which is necessary to obtain a product, good or service. It is therefore involves a "resource-driven" method.

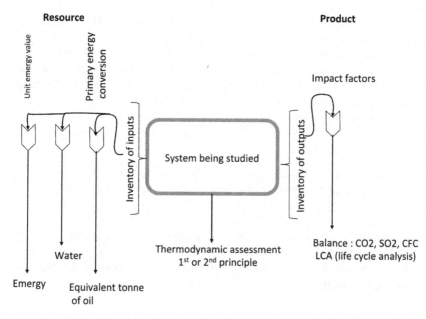

Figure I.1. *Synoptic "resources" or "product" analyses*

Developing the same comparative base between energy resources (whether fissile, fossil and renewable) makes it possible to construct a decision-making tool, which falls into a *sustainable development* optic.

Whatever the nature of energy resources (fissile, fossil or renewable), the latter is characterized by thermodynamic principles, the first of which expresses work and heat as two forms of one and the same notion, being internal energy or the thermodynamic function of a state of matter[1]. It is commonplace to say that internal energy's *weakness* arises from its adding together two quantities which lack the same quality. The works of S. Carnot (1796–1832) made it possible to demonstrate that producing work from a quantity of heat at a given temperature comes at a cost to Nature (this relating to the environment and being a loss of energy), thereby introducing Carnot's

---

1 A function of a given chemical's state is entirely distinguished by its changes in state and does not depend on the path that a chemical takes (through conversion) but only on its initial and final states. Thus, the internal energy of a non-reactive fluid depends, for example, on the pressure and the temperature, when subject to the parameters of its equation of state.

notion of efficiency[2], and the bases of the second principal. The functions of thermodynamic state (internal energy, enthalpy and other functions) are expressed in joules[3]. These functions do not depict the development of a given form since the notion of a path or route is clearly absent from this process. Thus a kilogram of coal is generally characterized by its low heating value, expressed in J/kg. However, it is the consequential result of substantial amounts of energy. Hence, there is a significant line of questioning by H.T. Odum (1924–2002), the biologist and Professor of the University of Florida, that may be expressed as:

> *Given that one joule of solar energy does not exert the same ecological footprint as one joule of fossil energy, how should this difference be accounted for?*

Professor H.T. Odum based his concept on the following observation: all earthly life[4] requires solar energy. He therefore took this energy as the reference point. Emergy returns to generalizing the concept of primary energy traditionally used to compare ultimate needs, indeed even engineering (2007):

– In Chapter 1, the fundamentals of emergy are set out. The original definition of emergy is put forward. Diagrams are introduced. The rules for emergy "propagation" within a diagram are clarified. The evaluation of various sources is then set out. The methodology of an emergetic analysis completes this chapter.

– In Chapter 2, five examples of applications are suggested, in line with European Union energy policy:

- wood energy is broached from the perspective of its emergy-CO_2 interaction;

- wind resources on a continental scale are then put forward;

---

2 Exergy takes into account the capacity to convert heat into work, relative to its environment, owing to a theoretical diathermy mechanism.

3 In 1824, Clément (1779–1841) introduced the calorie which, by convention, represents the energy which needs to be provided to raise a kilogram of water by one degree Celsius (or one degree Kelvin). As water has physical properties (in particular its heat capacity) which vary according to the temperature, a more stable fluid was selected – that is to say: air. In 1843, Joule (1818–1889) proposed taking as the reference point the quantity of heat which is needed to raise a kilogram of air by 1°C.

4 It is necessary to counterbalance this observation as life forms have recently been found close to submarine volcanoes.

- a comparison between thermal solar and photovoltaic systems is carried out;

- the process of producing biodiesel from palm oil is explained;

- an explanation of bioethanol production from seaweed then distinguishes macro-seaweed and micro-seaweed.

– Chapter 3 addresses the topic of recycling. By adopting an approach similar to the Lagrangian one to fluid mechanics, it is possible to establish that the intuitive notion of temporal independence of the specific emergy of materials necessitates slight variations. Discrete temporal equations are instituted according to the various configurations examined which are:

- the ideal "continuous" system without loss of matter within the entire "recycling" process;

- the "continuous" system with a loss of matter at the recycling level (rejection);

- the "continuous" system with loss of matter at the conversion level – notable applications relate, although not exclusively, to the iron and steel industry;

- the "discontinuous" system which mixes recycled products between 1 and n times.

In the final part, the loss of quality of matter is then envisaged. To conclude, an example of emergy with the loss of matter and quality is applied to aluminum:

– In Chapter 4, advanced concepts in the field of emergy are introduced. Having defined mineral emergy, the revised method is then proposed. The implications show, for example, that the specific emergy of a mineral is not a constant and depends upon the quality of the mine. The link with the previous chapter is then made, thereby showing the interaction between the production of a "raw" mineral and the recycling thereof.

The Fundamentals of Emergy

DEFINITION 1.1.–

> "*The emergy of a product (or a service) is defined as the amount of equivalent exergies, directly or indirectly required, for its production and takes the form of solar equivalent energy, denoted in seJ (solar equivalent energy)* [ODU 96]".

1.1. Concept

The definition of emergy should only draw a slight distinction between emergy and exergy. Nevertheless, one of the fundamental distinctions lies in the rules of calculation which are associated with emergy in a system, having inputs (emergy source) and outputs (the product or service):

Rule 1 – *all emergy sources which are mobilized to obtain a given output are allocated to such an output;*

Rule 2 – *co-products[1] of a process have the same total emergy;*

Rule 3 – *in the case of a split junction, the emergy of each branch is divided pro rata the energy passing through the latter; and*

Rule 4 – *emergy should not be measured twice:*

---

1 Co-products are products that stem from the same process at the same time but do not have the same chemical structure. For example:
– combustion engines (both internal and external) use fuel and produce work (which is generally converted in the form of electricity) and heat; and
– a sheep's meat and wool are both considered to be co-products.

- emergy recirculating within a given circuit should not be allocated twice,

- in the case of co-products joining together again, the emergy at such point should not exceed the emergy at the origin.

The original term for these rules is *emergy algebra*. Rule 4 induces Kirchhoff's laws which no longer apply when feedback is present. Indeed, the mathematical structure supporting the axiomatic framework of emergy is not algebra: this wording is imprecise and gives rise to rules which are incompatible [LEC 12].

1.2. Systemic emergy representation

The basic *"block diagram"* is used to describe emergy systems (see [ODU 95]). Nine main elements are described in Table 1.1.

As with all energy analysis, it is necessary to define the system studied beforehand, thus both the boundaries and the period have to be analyzed. The system is generally represented by the structure encompassing all conversions. Sources represented by circles are considered to be outside of the system. An emergy diagram then corresponds to descriptions of the interactions between the different stages of a system involved in the production of a product (whether a good or service). Figure 1.1 is a generic example of the emergy diagram.

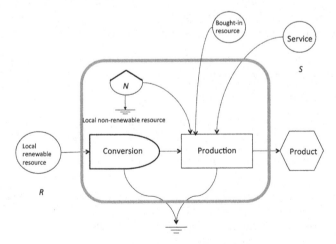

Figure 1.1. *Example emergy diagram*

Element	Symbol	Function	Example
Emergy connector			Cable, piping, or similar device
Emergy source			Renewable energy, oil or other source.
Environmental damage		Loss of energy (but not emergy)	Units considered to interact with the environment
Emergy conversion		Process stage	Distillation column and similar examples
Emergy conversion		Emergy production	Combustion engine, heat exchanger or others
Emergy junction			3-way valve
Storage/reservoir		Accumulator	Mines, natural gas fields and other energy source locations
Transaction		Block indicating a price (dotted line) for a given product	Interaction with an economic system
End consumer			Product: electricity, biodiesel, fish, automobile and others

Table 1.1. *Lexicon for block diagrams*

1.3. Application of the laws of emergy

The sources of an emergy diagram are independent of each other. It is possible to carry out the calculation on a source-by-source basis.

Product emergy is additive in relation to all sources.

Rule 2 indicates that the (total) emergy of co-products is equal to the input emergy Em^i (see Figure 1.2):

$$Em_{p1} = Em_{p2} = Em^i \qquad\qquad [1.1]$$

The emergy of co-products gains value from it(s) input(s). However, these emergies should not be double-counted (in the event of reunification).

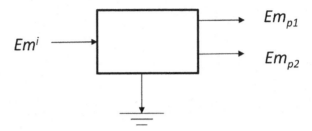

Figure 1.2. *Diagram of co-products*

Rule 3 indicates that the (total) input emergy within a junction is distributed in proportion to energy outputs. If we observe x_j energy fraction outputs from a distributor (see Figure 1.3), then the emergy of each branch of the distributor is obtained by:

$$Em_{pj} = x_j \, Em^i \qquad\qquad [1.2]$$

The specific emergy output of a distributor is equal to its input emergy.

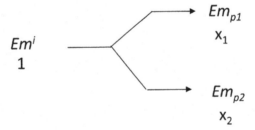

Figure 1.3. *Diagram of a distributor*

1.3.1. *Example 1: co-products-based system*

To refer to Figure 1.4, by way of example, the converter *T1* has two co-products. The first co-product is controlled by a distributor in respect of 2/5 to output *P1*, and as respects 2/5 to the converter *T2*. The *T2* output is also a *T4* input. The second *T1* co-product is also controlled by a distributor as respects 1/3 towards output *P3*, and as respects 2/3 towards the converter *T3*. The *T3* output is also a *T4* input. The *T4* output gives the output *P2*.

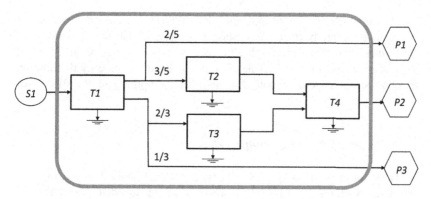

Figure 1.4. *Example with co-products (suggested by L. Truffet)*

– The *T1* co-products have the emergy value of the source *S1* (the application of the second rule).

– The emergy of *P1* is thus 2/5 of *S1* (applying the third rule).

– The *T2* input has a value of 3/5 of *S1*, and the output of *T2* is equal to its input (by applying the first rule).

– Likewise, the emergy of *P3* is equal to 1/3 of *S1* (by applying the third rule).

– The *T3* input is equal to 2/3 of *S1* and the *T3* output is equal to its input (by applying the first rule).

– The *T4* inputs stem from the co-products of the *T1* converter. The pathway from the source to the product, i.e. the route is here of primary importance so as to avoid double-counting. The (erroneous) application of the first rule might lead to adding up input emergies, thus: (3/5 + 2/3) of *S1*, i.e. 19/15 of *S1*. This result is evidently erroneous as it is superior to the *S1* emergy. When these co-products meet, rule four defines the means of calculation, indicating that there should be no double-counting. It is not the total inputs but the maximum inputs to be considered. The emergy of *P2* is equal to the greater of (3/5, 2/3) of *S1*, i.e. $Em_{p2} = 2/3 \, Em_{S1}$.

The fourth rule is difficult to implement. Light provides a possible analogy for its application. We might imagine that the co-products are the various colors of the spectrum which all have the same source. When they recombine, they reconstruct the light source in its original form (but do nothing further).

The enunciation of these four rules does not indicate the order in which it is necessary to apply them. Nevertheless, it appears that the application of these rules in reverse order to that in which they are stated appears to avoid the risk of double-counting. Rules 2 and 3 indicate the emergy distribution for two specific structures (both distribution and co-products). Rule 1 defines the other cases.

1.3.2. *Example 2: system with a distributor in a loop*

Take, for example, a system with two sources *S1* and *S2*, two converters and a distributor (see Figure 1.5). When reaching the distributor, the 3/5 proportion is returned to the converter *T1*.

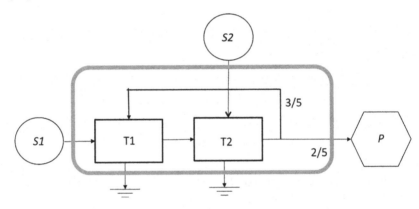

Figure 1.5. *System with a distributor in a loop*

1) Let us simply consider source *S1*, i.e. the emergy from source *S2* is zero, by applying the additivity property of emergy sources.

The T1 converter has two inputs: one comes directly from source *S1* and the other is the recirculation of this source within the system. Whatever the emergy value of this recirculating source, it is lower or equal to the value of the *S1* input. The emergy coming out of *T1* (output), which is denoted as Em_{T1}^O, is:

$$Em_{T1}^O = Em_{S1}$$

[1.3]

By applying the first rule, the observed emergy learning the *T2* converter Em_{T2}^o, is equal to the emergy entering (the input), denoted as Em_{T2}^i, the emergy itself coming from *T1*:

$$Em_{T2}^o = Em_{T2}^i = Em_{T1}^o = Em_{S1} \qquad [1.4]$$

By applying the third rule, the emergy in the distributor terminals is proportional to the quantity of energy. It is possible to arrive at the output emergy Em_p due to the source *S1*:

$$Em_p = \frac{2}{5} Em_{T2}^o = \frac{2}{5} Em_{S1} \qquad [1.5]$$

Nevertheless, if we apply the first rule, that is the emergies of inputs from the converter stage should be allocated upon its output, we obtain:

$$Em_p = Em_{S1} \qquad [1.6]$$

Thus the conflict between rules 1 and 4 becomes apparent.

We may note that emergy carried by the recirculation connection in the 3/5 proportion is $\frac{3}{5} Em_{T2}^o$, i.e. $\frac{3}{5} Em_{S1}$. The application of rule 4 prohibits consideration of this recirculation emergy with respect to *S1* for calculating input emergies: doing so may involve double-counting as the *S1* source emergy may have been double counted. Equation [1.4] is in fact:

$$Em_{T1}^o = \max\left(Em_{S1}; \frac{3}{5} Em_{S1}\right) = Em_{S1}$$

Again, the analogy with light is possible. The light from source *S1* enters *T1*, goes through *T2* then is distributed as a *T2* output. The 3/5 share channeled to *T1* is, in fact, contained within the source light flow: recirculation cannot produce more light than the initial source light.

2) Let us solely consider source *S2,* i.e. the value of the source *S1* is considered to be zero.

The emergy output from the *T2* converter is:

$$Em_{T2}^o = Em_{S2} \qquad [1.7]$$

This emergy is divided into 2/5 for the output and 3/5 passing within the recirculation loop:

$$Em_p = \frac{2}{5} Em_{T2}^o = \frac{2}{5} Em_{S2} \qquad [1.8]$$

The emergy output from the *T1* converter (coming from source *S2*) is:

$$Em^o_{T1} = \frac{3}{5}\ Em^o_{T2} = \frac{3}{5}\ Em_{S2} \tag{1.9}$$

On the other hand, at the *T2* converter level Em^o_{T1} should not be added to that coming from the source Em_{S2}: this would amount to double-counting.

The successive application both of the rules and the additive property of the various sources gives:

$$Em^o_{T1} = Em_{S1} + \frac{3}{5} Em_{S2}$$

$$Em^o_{T2} = Em_{S1} + Em_{S2}$$

$$Em_p = \frac{2}{5}\ (Em_{S1} + Em_{S2})$$

1.3.3. *Example 3: system with co-products, distributor and loop*

The example suggested is shown by Figure 1.6.

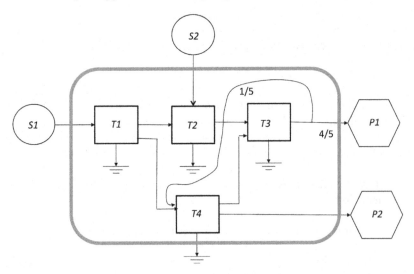

Figure 1.6. *System with co-products, a distributor and a loop*

Two co-products are obtained as outputs of the *T1* converter and from *T4*. A distributor is located at the *T3* converter output.

The additive property of the sources makes it possible to successively consider sources *S1* and then *S2*:

– calculating the emergy of output *P1* is somewhat similar to the previous example. There are several routes from the source *S1* source to the output *P1*;

– emergy from *S1* goes through *T1*, then *T2* and then *T3*. In leaving *T3*, a fraction (1/5) is sent to *T4*, a conversion which gives rise to two co-products, the first being redirected to *T3*. This share of emergy will not be taken into account so as to avoid double-counting. The share of emergy by this route from the source *S1* in the output *P1* therefore equals 4/5;

– the emergy from *S1* goes through *T1* then *T4* and *T3* and, after going through a distributor, ends up at the output *P1*.

By these two routes, the source *S1* contributes in exactly the same way to the output *P1*, i.e. max $(\frac{4}{5}Em_{S1}; \frac{4}{5}Em_{S1})$.

Likewise, the share of the source *S2* within the output also equals 4/5.

The emergy of the product *P1* is therefore:

$$Em_{p1} = \frac{4}{5}(Em_{S1} + Em_{S2}) \tag{1.10}$$

Calculating the emergy of product *P2* makes it possible to state the fourth rule during the presence of co-products. Again the additive property of sources is used.

In considering the source *S1*, there are:

– *T1* outputs for the emergy: $Em_{T1}^{o} = Em_{S1}$ (the second rule);

– the *T2* output which is equal to its input: $Em_{T2}^{o} = Em_{T1}^{o} = Em_{S1}$ (the first law);

– likewise, the output *T3* is: $Em_{T3}^{o} = Em_{T2}^{o} = Em_{T1}^{o} = Em_{S1}$ (the first rule);

– in leaving *T3*, the distributor passes the 1/5 fraction of *T3* to *T4*; and

– in terms of $T4$, there is another route which leaves the source $S1$ towards the converter $T4$. It is necessary to assess it so as to be able to supply the share of emergy leaving $T4$, which contributes to the product $P2$. The second co-product $T1$ is sent towards the converter $T4$. The emergy leaving the $T4$ converter is the maximal value coming from the source $S1$, that is this value does not result in any double-counting:

$$Em_{T4}^o = \max\left(\frac{1}{5} Em_{S1}; Em_{S1}\right) = Em_{S1} \qquad [1.11]$$

It is necessary to do likewise for the source $S2$. The $S2$ emergy circulates within the converter T. It is then sent to $T3$. The distributor directs 1/5 of the $S2$ energy to the converter $T4$. The emergy leaving $T4$ is therefore:

$$Em_{T4}^o = \frac{1}{5} Em_{S2} \qquad [1.12]$$

In the end we obtain the emergy of the product $P2$, being the sum of the respective source contributions:

$$Em_{P2} = Em_{S1} + \frac{1}{5} Em_{S2} \qquad [1.13]$$

The calculation of emergy corresponds to determining the routes of a source S towards a product P, using the analogy of the property of light to avoid double-counting.

1.3.4. Complete worked example

We now consider the system in Figure 1.7. There is a distributor after source $S2$; 3/10 of the source emergy is sent to the $T2$ converter and 7/10 to the $T3$ converter. The outputs $T2$ and $T3$ are inputs of $T4$. The $T4$ output is made up of co-products. One co-product is the product $P2$. The second co-product is sent to a distributor which divides the flow by two towards $T5$ and $T1$. Two co-products are outputs of $T5$. One of the co-products is an input of $T3$, and makes up a loop. The second co-product is sent to a distributor; the first feed, 1/2, constitutes the output $P1$, and the other 1/2 is a $T1$ input and makes up a second loop.

The method proposed to determine product emergy is "monitoring" the emergy leaving given sources, whilst avoiding double-counting. This technique is known as the "track summing method" [TEN 88], initially developed without co-products. It is necessary to determine all possible routes which leave a given source going to a given product, without passing the same converter twice.

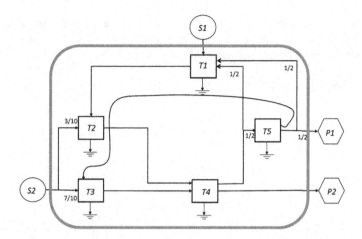

Figure 1.7. *Application of the rules of emergy to a case study*

Figure 1.8 shows the shortest route to reach output *P1* from source *S1*. As there is only one route, no double-counting is possible. The emergy of *P1* coming from *S1* depends solely on the possible distributions. The *T1* output is direct to *T2*, as well as from *T2* to *T4*, the latter being 1/2 connected to *T5*. To conclude, the *T5* output is 1/2 connected to *P1*. The contribution of *S1* to output emergy *P1* is:

$$Em_{P1} = \frac{1}{2}\frac{1}{2}Em_{S1} = \frac{1}{4}\,Em_{S1}$$

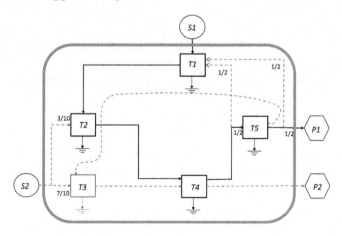

Figure 1.8. *The route from S1 to P1*

The shortest route from *S1* to *P2* is given in Figure 1.9. From source *S1*, it goes via *T1*, then by *T2*, then *T4* to arrive at *P2*. There is no distribution. The contribution from *S1* to the emergy of the product *P2* is:

$$Em_{P2} = Em_{S1}$$

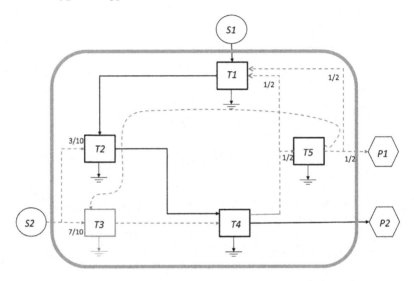

Figure 1.9. *The route from S1 to P2*

There are two routes from *S2* to *P1* involving no recirculation, see Figure 1.10;

– the route going from *S2* to *T3* then *T4* and *T5* reaches *P1*. This route has a distribution output from *S1* (7/10), a distribution output from *T4* (1/2) and a distribution output from *T5* of 1/2. The contribution of *S2* to emergy from *P1* by this route is:

$$Em_{P1} = \frac{7}{10}\frac{1}{2}\frac{1}{2} Em_{S2}$$

– the route going from *S2* to *T2* then *T4* and *T5* reaches *P1*. On this route, there is a distribution output from *S1* (3/10), a distribution output from *T4* (1/2) and a distribution output from *T5* (1/2). The contribution of *S2* to emergy from *P1* by this route is:

$$Em_{P1} = \frac{3}{10}\frac{1}{2}\frac{1}{2} Em_{S2}$$

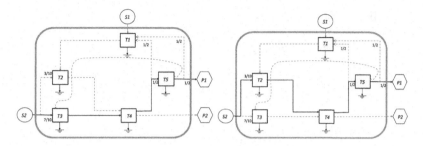

Figure 1.10. *Routes from S2 to P1*

These two routes stem from a distributor located after the source; it is thus necessary to add the contributions of *S2*. The contribution of *S2* to the emergy of *P1* is:

$$Em_{P1} = \left(\frac{7}{10}\frac{1}{2}\frac{1}{2} + \frac{3}{10}\frac{1}{2}\frac{1}{2}\right) Em_{S2} = \frac{1}{4} Em_{S2}$$

There are two routes from *S2* to *P2* involving no recirculation (see Figure 1.11);

– the route leaving from *S2* to *T3*, then *T4,* arrives at *P2*. There is an output distributor from *S2*. The contribution of *S2* to the emergy of *P2* is:

$$Em_{P2} = \frac{7}{10} Em_{S2}$$

– the route leaving from *S2* to *T2*, then *T4,* arrives at *P2*. There is an output distributor from *S2*. The contribution of *S2* to the emergy of *P2* is:

$$Em_{P2} = \frac{3}{10} Em_{S2}$$

These two routes stem from a distributor located after the source; it is thus necessary to add the contributions of *S2*. The contribution of *S2* to the emergy of *P2* is:

$$Em_{P2} = \left(\frac{7}{10} + \frac{3}{10}\right) Em_{S2} = Em_{S2}$$

For the *P1* output this gives:

$$Em_{P1} = \frac{1}{4}(Em_{S1} + Em_{S2})$$

For the *P2* output this gives:

$$Em_{P2} = Em_{S1} + Em_{S2}$$

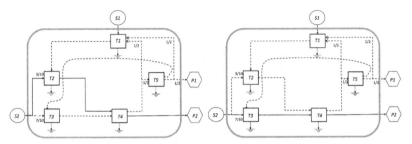

Figure 1.11. *Routes from S2 to P2*

1.3.5. *Exercise*

Li *et al.* [LI 10] proposed the diagram in Figure 1.12.

It proves that:

$$Em_{P1} = \frac{2}{3} Em_{S1} + \left(\frac{5}{8} + \frac{3}{8}\frac{4}{5}\right)\frac{1}{3}\frac{2}{3} Em_{S2}$$

$$Em_{P2} = \frac{1}{3}\frac{4}{5} Em_{S1} + \left(\frac{5}{8}\frac{4}{5} + \frac{3}{8}\frac{4}{5}\frac{1}{3}\frac{1}{3}\frac{4}{5}\right) Em_{S2}$$

$$Em_{P3} = \frac{1}{3}\max\left(\frac{1}{5};\frac{2}{3}\right)Em_{S1} + \left(\frac{5}{8}\max\left(\frac{1}{5};\frac{2}{3}\right) + \frac{3}{8}\frac{4}{5}\left(\frac{2}{3} + \frac{1}{3}\frac{1}{3}\max\frac{2}{3} + \frac{1}{3}\frac{1}{3}\max\left(\frac{1}{5};\frac{2}{3}\right)\right)\right)Em_{S2}$$

$$Em_{P4} = \frac{3}{8}\frac{1}{5} Em_{S2}$$

Hint: pay attention to the route passing through $4 - 6 - 1 - 3 - 5$.

1.4. Source emergy

To initiate an emergy analysis, inputs of the system being analyzed should be converted into emergy. The assessment of the geobiosphere makes it possible to define tidal emergies and sources of natural radioactivity. Fossil resources (coal, natural gas and oil) from different geological eras are analyzed. Minerals are assessed from a so-called (natural) enrichment factor. For interactions between the studied system and the economic environment, the emergy value of a given currency is needed.

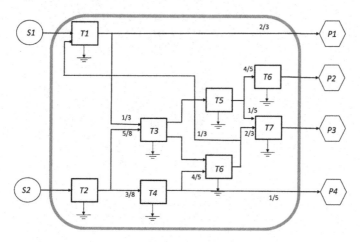

Figure 1.12. *Example of the application of the rules of emergy*

DEFINITION 1.2.– *Transformity is defined as the relationship between emergy (sources) with product exergy (seJ/J).*

Transformity is the inverse of an integrated output within given timescales, which are linked not to conversion but to the source in question.

By convention, transformity linked to solar radiation Tr is used as a point of reference: $Tr_S = 1$ seJ/J.

By extension, the unit emergy value (UEV) is defined as the emergy relationship for a product unit (for example: seJ/g, seJ/km). Transformity is a given unit emergy value.

1.4.1. *The basic geobiosphere paradigm*

The original paradigm [ODU 00] considers that exchanges within the geobiosphere group together four principal sources: the sun, the effects of gravitational forces of the moon mainly on the oceans, heat dissipated from the Earth's core, and lastly, natural sources of radioactivity dispersed within the crust (see Figure 1.13).

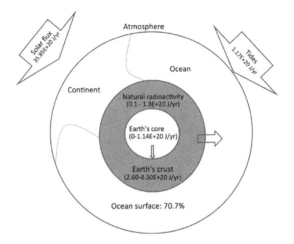

Figure 1.13. *The representation of emergy exchanges within the geobiosphere*

Brown and Ulgiati [BRO 10] described and updated fundamental elements based on calculations. By considering the Earth's crust and the oceans as two co-products[2], it is possible to establish the following annual emergy equations:

– the annual emergy of the Earth's crust is equal to the emergy received from the Sun, the Moon (in the form of tides), sources of natural radioactivity and the Earth's core over the period in question:

$$Em_{Cr} = Em_S + Em_M + Em_{rad} + Em_{ec} \qquad [1.14]$$

– the emergy of the ocean is equal to the emergy received from the Sun, the Moon and the Earth's crust:

$$Em_O = Em_S + Em_M + Em_{Cr-O} \qquad [1.15]$$

In respect of the emergy coming from the Earth's crust, the emergies of the Sun and the Moon should not be double counted. The exergy flow coming

---

2 In Brown and Ulgiati's article [BRO 10], it is not explicitly stated that the Earth's crust and the ocean are co-products. However, the emergy that they receive from the sun and the tide is the same (without weightings related to the surfaces occupied), so they can be labelled co-products.

from the crust is assumed to be proportional to the ocean surface (70.7% of the planet, denoted as %O):

$$Em_{ec-o} = \%O \ Em_{Cr} \qquad [1.16]$$

As the ocean and the crust are considered to be co-products, the emergy of the ocean is expressed as:

$$Em_O = Em_S + Em_M + \%O \ Em_{Cr} \qquad [1.17]$$

Emergy rule 4 excludes double counting in the event of sources recombining. It is necessary to include the maximum value of the source in question. In the case of oceanic emergy, we thus obtain:

$$Em_O = max\left(Em_S; \%OEm_S\right) + $$
$$max\left(Em_M; \%OEm_M\right) + \%O\left(Em_{rad} + Em_{ec}\right) \qquad [1.18]$$

$$Em_O = Em_S + Em_M + \%O \ (Em_{rad} + Em_{ec}) \qquad [1.19]$$

Equations [1.14] and [1.19] may thus be rewritten by introducing associated transformities:

$$\begin{cases} Tr_{cr}Ex_{cr} = Tr_S Ex_S + Tr_M Ex_M + Tr_{rad}Ex_{rad} + Tr_{ec}Ex_{ec} \\ Tr_O Ex_O = Tr_S Ex_S + Tr_M Ex_M + \%O \ (Tr_{rad}Ex_{rad} + Tr_{ec}Ex_{ec}) \end{cases} \qquad [1.20]$$

With: $Tr_S = 1 \ seJ/J$.

Two assumptions are introduced. Firstly, ocean transformity and the tides (the Moon) are equal. Secondly the transformities of the Earth's core, natural radioactivity and the Earth's crust are equal. On the basis of these two assumptions, we arrive at:

$$\begin{cases} Tr_O = \dfrac{Ex_{cr} + (1-\%O)(Ex_{rad} + Ex_{ec})}{Ex_O} \ Tr_{cr} \\ Tr_{cr} = \dfrac{Ex_S}{(Ex_{cr} - Ex_{rad} - Ex_{ec})\left(1 - \frac{Ex_M}{Ex_O}\right) - \frac{Ex_M}{Ex_O}(Ex_{rad} + Ex_{ec})\%O} \ Tr_S \end{cases} \qquad [1.21]$$

Brown and Ulgiati [BRO 10] retained the constant flow of the geobiosphere (see Table 1.2). The transformities calculated are indicated in Table 1.3.

It should be noted that:

– for the present calculations, the exergy of the crust is considered to be 8.50E20 J/year, that of the core 1.15E20 J/year and that of natural radioactivity 1.07E20 J/year. This particular choice is in the defined interval stated in Table 1.2;

– the value of 15.2E+24 seJ/year is the annual emergy value of the geobiosphere which is used in works in the field [BRO 10]. Before 2010, the value calculated by Odum [ODU 96] was 15.83E+24 seJ/year. Proportionality makes it possible to complete reference value calculations. The model developed in equation [1.21] is not that of the bibliography [BRO 10]. The values published are 72,400 seJ/J for the tides and 20,300 seJ/J for the crust. These values are greater than those given in Table 1.3 as the original work executed a double counting exercise.

Raugei [RAU 12] suggested another approach to calculating geobiospherical emergy. He introduced a 3D vector translating the respective source contributions. Thus solar transformity is written [1, 0, 0] seJ/J, tidal transformity [0, 1, 0] seJ/J and that of the Earth's crust [0, 0, 1] seJ/J. This premise avoids having a geobiospherical model which is highly, or indeed too simplified.

	Power Conversion into power/year	Exergy 1.00E+20 J/year
Solar	113,700	35,850
Tide	3.7	1.17
Earth's crust	8.1 – 27.0	2.6 – 8.5
Radioactivity	0.3 – 4.1	0.1 – 1.3
Heat of earth's core	0 – 4.4	0 – 1.4

Table 1.2. *Resources intervening in the geobiosphere, see Brown and Ulgiati [BRO 10], consistent with Hermann [HER 06]*

	Transformity seJ/J	Emergy 1.00E+24 seJ/year
Solar flux	1	3.6
Sources of terrestrial heat	**18,000**	**4.00**
Tidal energy	**66,000**	**7.72**
Emergy of the geobiosphere		**15.2**

Table 1.3. *Transformity of geobiosphere fluxes*

1.4.2. *Fossil resources*

1.4.2.1. *Coal*

There are four categories of coal: anthracite, bituminous, lignite (brown coal) and sub-bituminous coal. The deposits of the four families, categorized as "hard" (anthracite and bituminous) or "friable" (sub-bituminous and lignite), are summarized according to geological formation ages; see Table 1.4 taken from Brown *et al.* [BRO 11].

Geological period	% Coal	Quantity (E9 MT)		Available energy (E18 J)	
		Hard	Friable	Hard	Friable
Devonian	0.1	0.5	0.4	11.5	7.7
Carboniferous	24.3	116.9	103.7	2,806.4	1,866.8
Permian	31.7	152.5	135.3	3,661.0	2,435.3
Triassic	0.4	1.9	1.7	46.2	30.7
Jurassic	16.8	80.8	71.7	1,940.2	1,290.6
Cretaceous	13.3	64.0	56.8	1,536.0	1,021.8
Tertiary	13.5	65.0	57.6	1,559.1	1,037.1
Total	**100%**	**481.7**	**427.2**	**11,560.3**	**7,690.1**

Table 1.4. *Coal deposits categorized according to geological age [BRO 11]*

According to the various geological ages (e.g. Carboniferous, Jurassic, Cretaceous) decomposition of an amount (PF_1) of lush vegetation (receiving sunlight and interacting with its environment) has resulted in the formation of peat. A proportion of this peat (PF_2) has remained close to the Earth's mantle and has become coal over time. Figure 1.14 illustrates the main stages of the coal production process from an emergy point of view.

In order to estimate the emergy content of coal according to its given category, the methodology employed by Brown *et al.* [BRO 11] determines the total input emergies in the overall process.

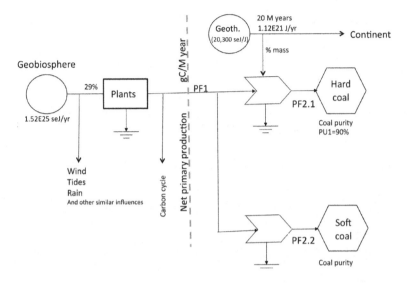

Figure 1.14. *The coal formation process*

The coal emergy unit value, whether hard or friable coal, comes from two sources – the geobiosphere for plants and the Earth's crust for carbonification. As the geological period linked to the vegetation is necessarily earlier than the time of coal formation, the sources are independent in time and the property of source additivity applies.

$$UEV(gCcoal) = UEV(gCcoal, Geo) + UEV(gCcoal, Cr)$$

For the unit emergy value of plants, Brown *et al.* [BRO 11] rely on the works of Beerling [BEE 99] and state the net primary production of carbon from vegetation according to geological eras *NPP* (*era*) (see Table 1.5). The unit in the first column is a gram of carbon for every million years. This vegetation receives part of the annual emergy of the geobiosphere $(Em(Geo), 15.2E24$ seJ/year – see section 1.4.1). The proportion indicated is 29%. The unit emergy value for a gram of carbon in plants is therefore:

$$UEV(gCplant, era) = \frac{Em(Geo)*Pg *29\%}{NPP(era)}$$

Pg (geological period) corresponds to one million years so as to be homogenous to the unit of net production. The value of the numerator is 4.41E+30 seJ/My.

The second column of Table 1.5 corresponds to the unit emergy value of a gram of carbon in plants. The low heating value of organic carbon is 45.7 kJ/g C [SAL 76]. The unit emergy value of a joule of carbon in plants is indicated in the third column.

Geological period	Vegetation *NPP* (gC/My)	Unit emergy seJ/gC	*UEV* seJ/J
Devonian	4.50E+22	9.80E+07	2.14E+03
Carboniferous	4.23E+22	1.04E+08	2.28E+03
Permian	5.40E+22	8.16E+07	1.79E+03
Triassic	5.40E+22	8.16E+07	1.79E+03
Jurassic	1.60E+23	2.76E+07	6.03E+02
Cretaceous	1.39E+23	3.17E+07	6.94E+02
Tertiary	1.32E+23	3.34E+07	7.31E+02

Table 1.5. *Transformity of vegetation according to geological period [BRO 11]*

The mass assessment allows us to calculate the vegetation required to obtain a gram of coal, from a carbon purity PU. One gram of coal requires 1/PF2 gram of peat. One gram of peat requires 1/PF1 gram of vegetation. The contribution of the geobiosphere to the unit emergy value of gram of coal is therefore:

$$UEV(gC\ coal, Geo) = \frac{1}{PF1} \frac{1}{PF2} PU \frac{29\%}{NPP(era)} Pg\ Em(Geo)$$

An estimate of the factors of the *FP* process is given according to the type of coal (hard or friable); see Table 1.6 and the works of Duke [DUK 03]. Nevertheless, there is a wide disparity in the values within published works.

Type of coal	Carbon purity (%)	*PF₁* (Process Factor 1): Peat-based vegetation	*PF₂* (Process Factor 2): Peat in coal
Hard	90%	7%	69%
Friable	70%	7%	95.6%

Table 1.6. *Process factor 1 – peat-based vegetation, Process factor 2 – peat in carbon [BRO 11]*

The contribution of the Earth's crust to the unit emergy value of a gram of carbon within coal corresponds to an equal mass division of the emergy of the crust for the period of carbonification, denoted as A_{carb}, estimated at 20 million years:

$$UEV(gC\ coal, Cr) = \frac{1\ gC}{m_{cr}}\ Ex_{cr}\ Tr_{cr}\ Pg_{carb}$$

where m_{cr} is the mass of the crust (2.82E+25 g), Ex_{cr} is the annual exergy of the Earth's crust (8.51E+20 seJ/year [BRO 11]) and the transformity of the crust has a value attributed by published works of 20,300 seJ/J. The value is 1.55E+07 seJ/g.

Transformities differ according to geological ages (see Table 1.7). The values of Table 1.6 differ, of the order of 10%, from those of the works of Brown et al. [BRO 11], resulting from a Monte-Carlo type approach.

Geological period	Transformity of hard coal (seJ/J)	Transformity of friable coal (seJ/J)
Devonian	8.32E+04	5.99E+04
Carboniferous	8.85E+04	6.37E+04
Permian	6.94E+04	4.99E+04
Triassic	6.94E+04	4.99E+04
Jurassic	2.36E+04	1.69E+04
Cretaceous	2.71E+04	1.94E+04
Tertiary	2.85E+04	2.05E+04

Table 1.7. *Transformity of coal according to geological period [BRO 11]*

1.4.2.2. Oil and natural gas

A similar approach for oil and gas deposits allows us to obtain the transformities of fossil resources. Table 1.8 affects offshore deposits and Table 1.9 onshore deposits [BRO 11].

Geological period	Transformity of off-shore oil (seJ/J)	Transformity of off-shore natural gas (seJ/J)
Silurian	2.10E+05	1.70E+05
Devonian	4.50E+05	1.64E+05
Early Permian	2.18E+05	1.77E+05
Late Jurassic	6.28E+04	5.11E+04
Middle Cretaceous	8.41E+04	6.83E+04
Oligocene–Miocene	1.16E+05	9.41E+04
Average mass	**1.48E+05**	**1.71E+05**

Table 1.8. *The transformity of oil and off-shore natural gas*

Geological period	Transformity of terrestrial oil (seJ/j)	Transformity of terrestrial natural gas (seJ/J)
Early Permian		5.08E+05
Middle Cretaceous	2.34E+05	1.90E+05
Oligocene–Miocene	3.00E+05	2.43E+05

Table 1.9. *The transformity of oil and terrestrial natural gas*

1.4.3. *Minerals*

DEFINITION 1.3.– *The exploitation factor (EF) of a mineral in a mine corresponds to the economically exploitable concentration of the latter related to its average proportion on Earth.*

The level of quality of the exploitation of a mine corresponds to the proportion of the mineral in relation to its economic interest.

The most common 50 minerals are given in Table A.1. Figure 1.15 shows these minerals according to their average concentration on Earth.

The enrichment factor of a mineral EF_m is defined by:

$$FE_m = \frac{MPT_m}{\%m_m}$$

with MPT_m being the minimal profitability threshold and $\%m_m$ the mass percentage of the given mineral in the Earth's crust.

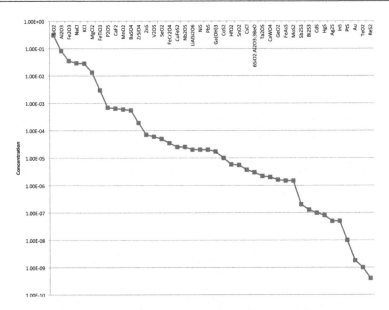

Figure 1.15. *Concentration of the major
minerals (see Appendix, Table A.1)*

Figure 1.16 shows that the rarer a mineral is (i.e. having a lower concentration) the greater its required concentration for exploitation.

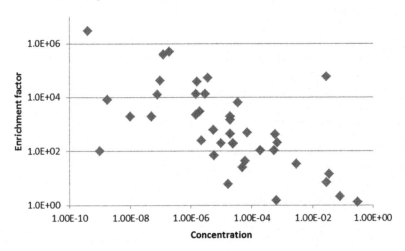

Figure 1.16. *Enrichment factor (for high quality
minerals) according to concentration*

By way of example, Table 1.10 offers the enrichment factor according to estimated profitability threshold.

	% mass in the crust	Minimum profitability threshold (MPT)	Enrichment factor (EF)
Aluminum	8.00E-02	17%	2.13E+00
Iron	3.50E-02	50%	1.43E+01
Titanium	3.00E-03	10.00%	3.33E+01
Manganese	6.00E-04	25%	4.17E+02
Chromium	3.50E-05	23%	6.57E+03
Nickel	2.00E-05	0.90%	4.50E+02
Copper	2.50E-05	1%	2.00E+02
Lead	2.00E-05	4.00%	2.00E+03
Silver	5.00E-04	0.01%	1.60E-01
Mercury	8.00E-08	0.10%	1.25E+04
Gold	1.80E-09	1.50E-03	8.33E+05

Table 1.10. *Enrichment factor for some minerals*

Brown [BRO 13] and Cohen [COH 07] have proposed calculating the unit emergy value of a mineral as being the product of the enrichment factor and the unit energy value of the Earth's crust, or:

$$em_m = EF_m \, em_{cr} \qquad\qquad [1.22]$$

and

$$em_{cr} = \frac{Pg_{cr} \, Em_{cr}}{m_{cr}}$$

where the emergy of the crust Em_{cr} is defined by equation [1.14]. Time Pg_{cr} corresponds to the geological time to renew the Earth's crust, considered to be 250 million years. The mass of the Earth's crust is considered to be 2.82E+25 g.

Table 1.11 gives the unit emergy value (seJ/mineral g) for various minerals. In the Appendix, Table A.2 gives the unit emergy value of about 50 compounds. Silver distinguishes itself from the linear trend in the graph of Figure 1.17.

	Unit emergy value seJ/g
Earth's crust	**1.35E+08**
Aluminum	2.86E+08
Iron	1.93E+09
Titanium	4.49E+09
Manganese	5.61E+10
Chromium	8.86E+11
Nickel	6.06E+10
Copper	2.70E+10
Lead	2.70E+11
Silver	2.16E+07
Mercury	1.68E+12
Gold	1.12E+14

Table 1.11. *The unit emergy value of various minerals*

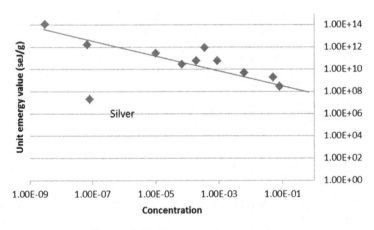

Figure 1.17. *Emergy value according to concentration*

1.4.4. EmCurrency

A country's currency may be calculated on an annual basis according to input flows. A database was developed by an organization (NEAD) in conjunction with the UN Environment Programme (2012). A country's gross domestic product (GDP) is obtained from inputs (see Figure 1.18). Relevant renewable resources are the Sun, the Earth's crust, the tides, the wind, waves and water. The source F represents fossil and fissile energies, minerals and materials. The source *Service* represents services imported into a country. The source N represents, for example, emergy from non-renewable extraction for forest management and similar uses. The source $N2$ corresponds to exploiting non-renewable resources without any conversion within the country.

The unit emergy value of a given country's currency is defined as the relationship between total inputs (converted to emergy) and the GDP in the country concerned. As inputs may vary annually, the unit emergy value of a currency depends on the given year. This approach [COH 12] makes it possible to connect the entire country macroscopically.

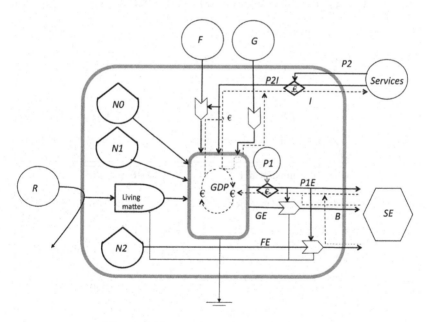

Figure 1.18. *Macro-economic perspective of a country (see [COH 12])*

Notation	Description	Unit	Calculation	Value
	Area of France	m²	Statistical	5.50E+11
	French population	#	Statistical	6.30E+07
GDP	Gross Domestic Product (PIB)	$	Statistical	2.90E+12
Em_R	Renewable energy	seJ	=max (1,2,3,4,6)+5 (*)	3.51E+23
Em_N	Non-renewable energy	seJ	=SOMME(15 - 23)	1.77E+23
Em_{N0}	Dispersed non-renewable production	seJ	=SOMME(15 - 18)	9.05E+21
$N1+N2$	Concentrated non-renewable production	seJ	=SOMME(19 - 23)	1.68E+23
Em_{N1}	Use of non-renewable energy	seJ	=(N1+N2)-N2	1.23E+23
$Em_{F(i)}$	Fuel, minerals and metals imports	seJ	=24+25+26	2.04E+24
$Em_{G(i)}$	Goods and electricity imports	seJ	=SOMME(27 - 34)	1.14E+24
Em_I	Financial import flow	$	= Qtt(35)	6.90E+11
Em_{P2i}	Imported services	seJ	=35	1.87E+24
Em_{N2}	Unused non-renewable energy exports	seJ	N2(f) - N2(m)	4.52E+22
$Em_{N2(f)}$	Unused fuel exports	seJ	=24% (36) (**)	1.55E+22
$Em_{N2(m)}$	Unused mineral and metal exports	seJ	=10% (37 - 38) (**)	4.62E+22
Em_{Fe}	Exported fuel, mineral and metal exports	seJ	=SOMME(36 - 38)	6.03E+23
Em_{Ge}	Exported goods and electricity	seJ	=SOMME(39 - 46)	9.85E+23
Em_E	Financial flow exports	$	= Qtt(47)	5.90E+11
Em_{P1e}	Service exports	seJ	=47	1.15E+24
Em_{P1}	Unit energy value of a euro in France (*EVEF*)	seJ/$	Total emergy/GDP	1.91E+12

Table 1.12. *Macro-emergy data for France (2008)*

This approach infers that the unit emergy value of a euro varies according to the area (scale) upon which it is calculated.

On the NEAD site, there is data for 180 countries. France's data is given in Appendix Table A.3. Table 1.12 is the synthesis of this data[3]. The "Calculation" column of this table refers to the line of Table A.3.

The more developed a country's economy and the more sophisticated its technology, the less recourse it has to the environment to create value. Conversely, the more a country exploits underground resources or forests, the greater the developmental burden upon nature, and therefore the more important the emergy value of its currency. Figure 1.19 illustrates this aspect very well; the two countries with the lowest currency emergy value were Japan and South Korea (in 2008).

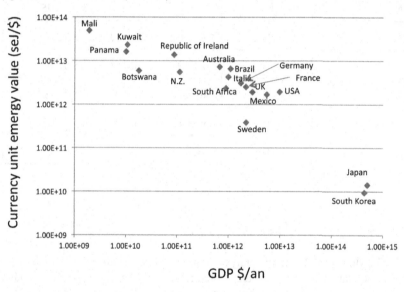

Figure 1.19. *Unit emergy value (in US$) according to countries' GDP (2008)*

1.4.5. *From fossil resources to fuel products*

Apart from the actual transformity of the raw resource, additional emergy must be added for its drilling/mining, transport and refining (see Table 1.13 [BRO 11]). This emergy provision is generally considered as a service

---

3 (*) Renewable emergy is calculated as the maximum between the sun, the tides and the wind. The initial calculation of emergy of the geobiosphere (see section 1.4.1) infers that these renewable energies are co-products and cannot be counted twice.
(**) These percentages have been used on a standardized basis.

coming from the economic system and is therefore expressed in emCurrency. Table 1.9 lists the values considered in published studies for fossil fuels.

Fossil resources	Exploitation Mining/boring (seJ/J)	Refining (seJ/J)	Transport (seJ/J)	Product transformity (seJ/J)
Soft coal	1.54E+04		2.97E+04	1.11E+05
Hard coal	1.23E+04		2.23E+04	1.32E+05
Natural gas	8.07E+03		–	1.78E+05
Petrol	8.07E+03	2.96E+04	1.26E+03	1.87E+05
Kerosene	8.07E+03	2.68E+04	1.26E+03	1.84E+05
Diesel	8.07E+03	2.40+04	1.26E+03	1.81E+05
LPG	8.07E+03	1.30+04	1.26E+03	1.70E+05
Petroleum residue	8.07E+03	1.54E+04	1.26E+03	1.73E+05

Table 1.13. *The transformity of fossil products*

1.5. Methodology of emergy analysis

The first step of emergy analysis requires gathering all inputs which cross the boundaries of the given system; see Figure 1.20 requires. This information is generally presented in the form of a table. Each constituent is referenced by an identifier, a designation, an associated unit, the quantity of the constituent used and the unit emergy content. The last column calculates the total emergy of the constituent (produced from the product quantity by its unit emergy content).

We may distinguish local resources (renewable or non-renewable) from those which are, by necessity, imported or transported within the market-based system.

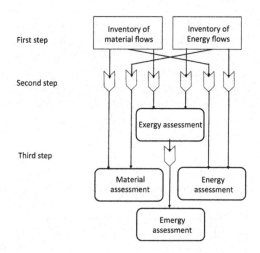

Figure 1.20. *Synoptic diagram of an emergy analysis (inspired by [BAR 04])*

1.5.1. *Example: production of aluminum in the United States*

Let us now turn to the production of aluminum, extracted from a mine, refined and converted into an ingot. Assume, as a means of simplifying these processes, that electricity is the only given source of energy. The emergy diagram associated with this process is given in Figure 1.21.

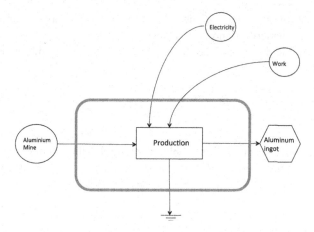

Figure 1.21. *Emergy diagram for the production of aluminum ingots*

In the case of US production, Buranakarn [BUR 98] indicated that a quantity of 4.17E+5 tons of minerals were used. From a database, he obtained the unit emergy content (1.17E+10 seJ/g), and accordingly the total emergy. By doing the same for electricity and human labor, he was able to deduce the total quantity of emergy. Knowing the annual production of aluminum ingots, he was able to calculate the unit emergy content of this product. This information can then enhance the database; Table 1.14 summarizes the above approach.

	Element	Unit/year	Quantity	UEV seJ/unit	Emergy seJ/year
1	Aluminum mineral	g	4.17E+11	1.17E+10	4.88E+21
2	Electricity	J	1.08E+15	1.74E+05	1.88E+20
3	Labor	$	2.09E+07	1.15E+12	2.40E+19
	Annual quantity	g	**4.00E+11**	**1.27E+10**	**5.08E+21**

Table 1.14. *Emergy table for annual production of USA aluminum (see [BUR 98])*

In this example, rule 1 applies. The other rules are not triggered. Thus described, the emergy approach is fairly simple. However, it is necessary to establish the basis and modify specific emergies to the given basic elements.

1.6. Emergy ratio

Emergy analysis makes it possible to distinguish three types of resources: (local) renewable resources R, (local) non-renewable resources, N, and flows passing through the economic system S (see Figure 1.22).

Five ratios have been defined, with the objective of making it possible to compare production solutions.

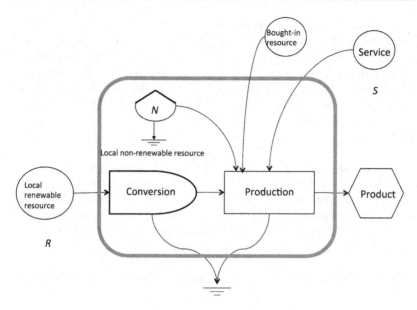

Figure 1.22. *Global production diagram*

Wilfart *et al.* [WIL 12] have synthesized, in an instructive way, the sources in Table 1.15.

	Description	Unit UEV	Notation
Environmental contribution (local)			
Renewable resource	Sun, wind, tidal, and other natural sources	seJ/J	R
Non-renewable resources	Fossil fuel, minerals and other fuels	seJ/unit	NR
Interaction with the economic system (external to a given system):			
Equipment (good)	Machine	seJ/$	S
Service	Electricity, employees labor	seJ/$	S
Input emergy (modified to outputs)			=R+NR+S

Table 1.15. *Classification of sources*

The classification of renewable resources is well defined as the boundaries of the system allow us to determine whether a resource passes through the economic system. There remain many different interpretations when modifying a resource: the types of resource, for example "non-renewable" or "service", depend on the given modeler's choice. Within the services, it is also possible to distinguish the "renewable" portion, even if it is not local. Whatever the choices made in the analysis, they should be explicit.

We may rapidly think that an eco-conception would consist in seeking the completion of a product, good or service with the lowest unit emergy value possible. Thus, two systems with the same production process may have different transformities; we might choose $TR1 < TR2$. If the second system uses more renewable resources for its production, then even if its transformity is higher than that of the first system, the second system should be retained.

To compare two systems, five main emergy ratios have been defined:

– the production ratio Emergy Yield Ratio (EYR) is defined as:

$$EYR = \frac{Em_P}{Em_S} \qquad [1.23]$$

The larger the EYR to the product iso-emergy, the lower the service contribution;

– the Emergy Loading Ratio (ELR) is defined by:

$$ELR = \frac{Em_F + Em_N}{Em_R} \qquad [1.24]$$

The smaller the ELR, the more significant the emergy portion of local renewable resources. Conversely, an increased value translates into a production system which mainly uses non-renewable resources or comes from an economic system;

– the Emergy Investment Ratio (EIR) is defined by:

$$EIR = \frac{Em_S}{Em_R + Em_N} \qquad [1.25]$$

With the product iso-emergy, the lower the EIR, the more the product will be obtained from local resources (whether renewable or non-renewable) and the more the emergy cost (investment) of flows passing through the economic system;

– the Emergy Sustainability Index (*ESI*) is defined as:

$$ESI = \frac{EYR}{ELR} \qquad\qquad [1.26]$$

An increased *ESI* value corresponds to a production system which uses local renewable resources and is therefore considered to be sustainable;

– Wu *et al.* [WU 15] defined the recycling fraction %*R*:

$$\%R = \frac{Em_R + Em_{S_R}}{Em_p} \qquad\qquad [1.27]$$

This fraction takes account of the share of local renewable emergy, but also connects the renewable part coming from the economic system.

Emergy and Converting Renewable Energy

In the European Union's Horizon 2020 program, the climate–energy package rests upon three pillars:

– the reduction of CO_2 emissions by 20%;

– the reduction of energy consumption by 20%;

– attaining a 20% share of renewable energy as part of total energy consumption.

This chapter advances five examples of emergy consumption:

1) Incentive policies for the fuel wood sector (in substitution for fossil fuel through heat or co-combustion networks) have been developed. These incentives are justified by CO_2 reduction. What would be the result(s) of an emergy analysis compared to a carbon footprint analysis?

2) European wind power sources are then looked at from an emergy viewpoint. The coastal potential is thus highlighted.

3) An emergy comparison between a solar thermal plant and a photovoltaic plant.

4) The production of the co-products, biodiesel and glycerol, from palm oil is another example of its application.

5) An emergy analysis of macro and micro-algae supplements these examples.

2.1. Substituting natural gas for fuel wood

We may raise the following question about replacing old boilers – using one (or several) decentralized fossil fuels compared to centralized production using fuel wood:

> *Is there a finite distance for transporting wood fuel? If the answer is affirmative, what criteria would define this limit?*

In this section, analytical criteria are based on both emergy analysis and the assessment of CO_2 levels. The original work was reported by Jamali-Zghal *et al.* [JAM 13].

2.1.1. *Analysis of a wood burning heating plant*

The limits of the wood burning heating plant are defined in Figure 2.1. Wood is considered to be a renewable source. Diesel is considered to be a fuel for transporting wood, ash and workers. The construction of the heating plant and the network are considered to come from the economic system (and are accounted for through their actual cost). The factory operation is carried out by those operating the heating plant.

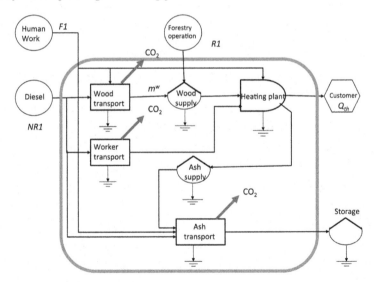

Figure 2.1. *Diagram of a wood-fueled heating plant*

The sources of Figure 2.1 classified, as per the description of section 1.6, under the following categories:

– *R1* wood resource emergy (renewable);

– *NR1* diesel emergy consumed transporting staff, wood and ash (non-renewable resources);

– *F1* emergy for service provision: work and investment.

2.1.1.1. *Emergy analysis*

Annual emergy of a wood boiler Em^{ws} is divided into six sources:

– the emergy from wood as a fuel, Em^w;

– the emergy to transport wood, $Em^{tr}(w)$;

– the emergy to transport ash, $Em^{tr}(ash)$;

– the emergy to transport workers, $Em^{tr}(e)$;

– the emergy linked to heating plant operation, $Em(e)$;

– the annual emergy linked to investments to construct the heating plant and the system, $Em^w(inv)$.

These sources are independent of each other which correspond to the application of Rule 1 (see section 1.1).

$$Em^{ws} = Em^w + Em^{tr}(w) + Em^{tr}(ash) + Em^{tr}(e) + Em(e) + Em^w(inv) \qquad [2.1]$$

The annual emergy of wood Em^w is the product of its annual mass m^w multiplied by its average calorific value LHV^w multiplied by its unit emergy value UEV^w:

$$Em^w = m^w \, LHV^w \, UEV^w \qquad [2.2]$$

The annual mass of fuel wood is the relationship between the energy generated Q_{th} (the system's demand for heat) and the product of the low heating value LHV^w and the average annual plant efficiency η^w.

$$m^w = \frac{Q_{th}}{LHV^w \, \eta^w} \qquad [2.3]$$

The low heating value of the wood varies according to humidity which is given as H:

$$LHV^w = LHV^w(0\%)\frac{(100-H)}{100} - 0.02443\,H \qquad [2.4]$$

The annual emergy for transporting wood $Em^{tr}(w)$ depends on the number of return trips Nb^w between the heating plant and the logging operation, the distance the wood is be transported D^w, the lorry's actual fuel consumption Cs^w, the diesel heating value LHV^d and the unit emergy value of diesel UEV^d. It is also necessary to take account of a no-load consumption factor γ^w as the lorry only actually transports wood when going to the heating plant.

$$Em^{tr}(w) = Nb^w\,D^w\,(1+\gamma^w)Cs^w LHV^d\,UEV^d \qquad [2.5]$$

The number of return trips Nb^w depends both on the lorry tonnage C^w and the quantity of wood to be transported m_w:

$$Nb^w = \frac{m_w}{C^w} \qquad [2.6]$$

The annual emergy for transporting ash $Em^{tr}(ash)$ is calculated in a similar way to that for transporting wood. It corresponds to the product between the number of return trips Nb^{ash} between the heating plant and the waste landfill facility, distance D^{ash}, actual lorry fuel consumption Cs^{ash} for carrying this, the diesel heating value and the UEV. A no-load consumption factor γ^{ash} is also introduced.

REMINDER.– Lorries transportating wood and ash are not necessary the same: their capacity can be different.

The annual emergy for the ash transport $Em^{tr}(ash)$ is expressed as:

$$Em^{tr}(ash) = Nb^{ash}\,D^{ash}\,(1+\gamma^{ash})Cs^{ash}LHV^d\,UEV^d \qquad [2.7]$$

We may obtain the number of return trips from the quantity of ash, obtained from the share of ash contained in the wood τ, and the lorry tonnage:

$$Nb^{ash} = \tau\frac{m_w}{C^{ash}} \qquad [2.8]$$

The annual emergy for carrying employees $Em^{tr}(e)$ is calculated by estimating the cumulative distance traveled from home to the heating plant D^e, actual car fuel consumption Cs^e, the diesel heating value and the UEV:

$$Em^{tr}(e) = D^e \, Cs^e LHV^d \, UEV^d \qquad [2.9]$$

Annual operating emergy for the heating plant is obtained from the number of running hours Nb^{tr} and the UEV linked to such operation – UEV^{tr}:

$$Em(e) = Nb^{tr} \, UEV^{tr} \qquad [2.10]$$

The annual construction emergy is obtained from the investment cost Inv^w and the unit emergy value of the euro $UEV^€$:

$$Em^w(inv) = Inv^w \, UEV^€ \qquad [2.11]$$

2.1.1.2. Assessment of CO_2 levels

Wood fuel is considered to be a renewable resource. Its associated carbon cycle is assumed to be neutral. The actual burning of wood within the heating plant emits CO_2, which is captured by young trees. The annual heating plant CO_2 assessment CO_2^{ws} is calculated from the emissions linked to transport (wood, ash and workers, $CO_2^{tr}(w) + CO_2^{tr}(ash) + CO_2^{tr}(e)$) and upstream emissions linked to the wood – $CO_2^{up}(w)$. Indeed, even if wood burning is considered to be carbon-neutral, logging operations emit CO_2 during the harvesting process and are so-called "upstream" of fuel wood. The annual CO_2 assessment of the wood fuel heating system CO_2^{ws} is:

$$CO_2^{ws} = CO_2^{tr}(w) + CO_2^{tr}(ash) + CO_2^{tr}(e) + CO_2^{up}(w) \qquad [2.12]$$

The annual CO_2 assessment linked to wood transport depends on the number of return trips Nb^w, distance D^w, the no-load consumption factor γ^w, specific lorry fuel consumption Cs^w, and diesel emission, these being local factors (linked to consumption) DEF^d (domestic emission factor) and upstream (linked to overall infrastructure) $UFEF^d$ (upstream fuel emissions factor), diesel oxidation ε and adjusting the relationship of molar masses $\frac{M_{CO_2}}{M_C}$.

$$CO_2^{tr}(w) = Nb^w \, D^w \, (1 + \gamma^w) Cs^w (DEF^d + UFEF^d) \, \varepsilon \, \frac{M_{CO_2}}{M_C} \qquad [2.13]$$

The annual CO_2 assessment $CO_2^{tr}(ash)$ linked to the ash transport is similar to that of wood:

$$CO_2^{tr}(ash) = Nb^{ash} \, D^{ash} \, (1 + \gamma^{ash}) Cs^{ash} (DEF^d + UFEF^d) \, \varepsilon \, \frac{M_{CO_2}}{M_C} \quad [2.14]$$

The annual CO_2 assessment $CO_2^{tr}(e)$ depends on the distance traveled by workers, the particular car fuel consumption, and both local and upstream emissions factors, as well as oxidation and the weighting attributed to molar masses:

$$CO_2^{tr}(e) = D^e \, Cs^e (DEF^d + UFEF^d) \, \varepsilon \, \frac{M_{CO_2}}{M_C} \quad [2.15]$$

To conclude this section, the annual CO_2 assessment, $CO_2^{up}(w)$, linked to activity upstream of the wood fuel plant is arrived at from the annual mass of wood m^w, the low heating value of wood LHV^w and the upstream emissions factor of the wood $UFEF^w$:

$$CO_2^{up}(w) = m^w \, LHV^w \, UFEF^w$$

2.1.2. Analysis of a natural gas heating system

The natural gas heating system is considered to be simpler as it does not require any workers to operate. It is therefore considered to be a source for natural gas consumption and another source of investment in this type of heating system. If we wish to compare two technical solutions, we must carry out the same set of analyses upon the second solution. The demand for heat Q_{th} is therefore the same as the previous section in 2.1.1.2. The quantity of natural gas energy Q^{ng} is therefore the relationship between the need for heat and the efficiency of the central heating system considered η^{ng}:

$$Q^{ng} = \frac{Q_{th}}{\eta^{ng}} \quad [2.16]$$

2.1.2.1. Emergy analysis

Annual emergy of a natural gas heating system Em^{ngs} depends on the energy consumption multiplied by the UEV of natural gas UEV^{ng} and the UEV linked to the infrastructures of the network distributing the natural gas (differentiating the heating system which is distributing the latter to consumers) UEV^{ngs}, and to add to the annual emergy throughout the heating plant $Em^{ng}(inv)$. We obtain the annual emergy of a natural gas heating system Em^{ngs} by:

$$Em^{ngs} = (UEV^{ng} + UEV^{ngs}) \, Q^{ng} + Em^{ng}(inv) \qquad [2.17]$$

The annual emergy linked to investment is obtained as calculated previously, (see equation [2.11]):

$$Em^{ng}(inv) = Inv^{ng} \, UEV^{\epsilon} \qquad [2.18]$$

2.1.2.2. Assessment of CO_2

The annual assessment of CO_2 for a natural gas heating system CO_2^{ngs} is mainly due to natural gas consumption. It is constituted both from a local factor DEF^{gn} and from an upstream factor $UFEF^{gn}$. We thus obtain:

$$CO_2^{ngs} = (DEF^{gn} + UFEF^{gn}) \, Q^{ng} \qquad [2.19]$$

2.1.3. Eco-limit for operation

It is possible to define the maximum distance beyond which wood fuel cannot be transported. As long as the emergy of a wood fuel heating system Em^{ws} is lower than that of a natural gas heating system Em^{ngs}, thus from an emergy point of view, wood systems have a lesser impact than natural gas systems:

$$Em^{ws}(D) < Em^{ngs}$$

The limit – the maximum distance for transporting wood D_{Em}^{max} – is therefore defined when the two heating systems have the same emergy:

$$Em^{ngs} = Em^{ws}(D_{Em}^{max}) \qquad [2.20]$$

This maximum distance D_{Em}^{max} may be written as:

$$D_{Em}^{max} = \frac{Em^{ngs} - (Em^{w} + Em^{tr}(ash) + Em^{tr}(e) + Em(e) + Em^{w}(inv))}{Nb^{w} \, (1 + \gamma^{w}) Cs^{w} LHV^{d} \, UEV^{d}} \qquad [2.21]$$

Eco-operation lays down the following condition:

$$D_{Em}^{eco} < D_{Em}^{max} \qquad [2.22]$$

The reasoning which is valid for emergy is also completely viable for CO_2 emissions. Wood fuel heating is relevant if its CO_2 assessment is lower than natural gas.

$$CO_2^{ws} < CO_2^{ngs} \qquad [2.23]$$

The maximum distance to transport $D_{CO_2}^{max}$ wood is determined when the two heating systems emit the same quantities of CO_2:

$$CO_2^{ngs} = CO_2^{ws}(D_{CO_2}^{max}) \qquad [2.24]$$

This distance may be expressed as:

$$D_{CO_2}^{max} = \frac{CO_2^{ngs}-\left(CO_2^{tr}(ash)+CO_2^{tr}(e)+CO_2^{up}(w)\right)}{Nb^w\,(1+\gamma^w)Cs^w\left(DEF^d+UFEF^d\right)\varepsilon\,\frac{M_{CO_2}}{M_C}} \qquad [2.25]$$

As previously, it is possible to have an eco-operation if the distance $D_{CO_2}^{eco}$ is lower than the maximum distance:

$$D_{CO_2}^{eco} < D_{CO_2}^{max} \qquad [2.26]$$

The comparison between maximum distances D_{Em}^{max} and $D_{CO_2}^{max}$ makes it possible to compare emergy analyses to CO_2 assessment analyses:

$$\frac{D_{Em}^{max}}{D_{CO_2}^{max}} = \frac{\dfrac{Em^{ngs}-\left(Em^w+Em^{tr}(ash)+Em^{tr}(e)+Em(e)+Em^w(inv)\right)}{Nb^w\,(1+\gamma^w)Cs^w LHV^d\,UEV^d}}{\dfrac{CO_2^{ngs}-\left(CO_2^{tr}(ash)+CO_2^{tr}(e)+CO_2^{up}(w)\right)}{Nb^w\,(1+\gamma^w)Cs^w\left(DEF^d+UFEF^d\right)\varepsilon\,\frac{M_{CO_2}}{M_C}}}$$

The $\dfrac{D_{Em}^{max}}{D_{CO_2}^{max}}$ relationship may be expressed as:

$$\frac{D_{Em}^{max}}{D_{CO_2}^{max}} = K\,\frac{\eta^w-\eta_{Em}^{min}}{\eta^w-\eta_{CO_2}^{min}} \qquad [2.27]$$

with

$$K = \left(\frac{\left(DEF^d+UFEF^d\right)\varepsilon\,\frac{M_{CO_2}}{M_C}}{LHV^d\,UEV^d}\right)\left(\frac{Em^{ngs}-Em^{tr}(e)-Em(e)-Em^w(inv)}{CO_2^{ngs}-CO_2^{tr}(e)}\right) \qquad [2.28]$$

$$\eta_{Em}^{min} = Q_{th} \frac{\left(UEV^W + \frac{\tau}{LHV^W Cs^{ash}}D^{ash}\left(1+\gamma^{ash}\right)Cs^{ash}LHV^d\,UEV^d\right)}{Em^{ngs}-Em^{tr}(e)-Em(e)-Em^W(inv)} \qquad [2.29]$$

$$\eta_{CO_2}^{min} = Q_{th} \frac{\left(\frac{\tau}{LHV^W Cash}D^{ash}\left(1+\gamma^{ash}\right)C^{ash}\left(DEF^d+UFEF^d\right)\varepsilon\frac{M_{CO_2}}{M_C}+UFEF^d\right)}{CO_2^{gns}-CO_2^{tr}(e)} \qquad [2.30]$$

MATHEMATICAL PROOF.–

We have:

$$\frac{D_{Em}^{max}}{D_{CO_2}^{max}} = \frac{\dfrac{Em^{ngs}-\left(Em^W+Em^{tr}(ash)+Em^{tr}(e)+Em(e)+Em^W(inv)\right)}{LHV^d\,UEV^d}}{\dfrac{CO_2^{gns}-\left(CO_2^{tr}(ash)+CO_2^{tr}(e)+CO_2^{up}(w)\right)}{\left(DEF^d+UFEF^d\right)\varepsilon\dfrac{M_{CO_2}}{M_C}}}$$

We may highlight the formula $\dfrac{\left(DEF^d+UFEF^d\right)\varepsilon\frac{M_{CO_2}}{M_C}}{LHV^d\,UEV^d}$:

$$\frac{D_{Em}^{max}}{D_{CO_2}^{max}} = \left(\frac{\left(DEF^d+UFEF^d\right)\varepsilon\frac{M_{CO_2}}{M_C}}{LHV^d\,UEV^d}\right)\frac{Em^{ngs}-Em^{tr}(e)-Em(e)-Em^W(inv)-\left(Em^W+Em^{tr}(ash)\right)}{CO_2^{gns}-CO_2^{tr}(e)-\left(CO_2^{tr}(ash)+CO_2^{up}(w)\right)}$$

Then, we may isolate the formulaic terms which do not depend on transport, namely the relationship $\dfrac{Em^{ngs}-Em^{tr}(e)-Em(e)-Em^W(inv)}{CO_2^{gns}-CO_2^{tr}(e)}$:

$$\frac{D_{Em}^{max}}{D_{CO_2}^{max}} = \left(\frac{\left(DEF^d+UFEF^d\right)\varepsilon\frac{M_{CO_2}}{M_C}}{LHV^d\,UEV^d}\right)\left(\frac{Em^{ngs}-Em^{tr}(e)-Em(e)-Em^W(inv)}{CO_2^{ngs}-CO_2^{tr}(e)}\right)\frac{1-\dfrac{\left(Em^W+Em^{tr}(ash)\right)}{Em^{ngs}-Em^{tr}(e)-Em(e)-Em^W(inv)}}{1-\dfrac{\left(CO_2^{tr}(ash)+CO_2^{up}(w)\right)}{CO_2^{gns}-CO_2^{tr}(e)}}$$

The formula K is therefore obtained:

$$\frac{D_{Em}^{max}}{D_{CO_2}^{max}} = K\frac{1-\dfrac{\left(Em^W+Em^{tr}(ash)\right)}{Em^{ngs}-Em^{tr}(e)-Em(e)-Em^W(inv)}}{1-\dfrac{\left(CO_2^{tr}(ash)+CO_2^{up}(w)\right)}{CO_2^{ngs}-CO_2^{tr}(e)}}$$

By replacing wood fuel emergies Em^w and the transport of ash $Em^{tr}(ash)$ by their formulaic equation, we obtain:

$$\frac{D_{Em}^{max}}{D_{CO_2}^{max}} = K \frac{1 - \frac{Q_{th}}{\eta^w}\frac{\left(UEV^w + \frac{\tau}{LHV^w C^{ash}}D^{ash}\left(1+\gamma^{ash}\right)C_S^{ash}LHV^d\,UEV^d\right)}{Em^{ngs} - Em^{tr}(e) - Em(e) - Em^w(inv)}}{1 - \frac{Q_{th}}{\eta^w}\frac{\left(\frac{\tau}{LHV^w C^{ash}}D^{ash}\left(1+\gamma^{ash}\right)C_S^{ash}\left(DEF^d + UFEF^d\right)\varepsilon\frac{M_{CO_2}}{M_C} + UFEF^w\right)}{CO_2^{ngs} - CO_2^{tr}(e)}}$$

At the beginning of this chapter, the climate and energy package in the Horizon 2020 plan was mentioned. It is also entirely possible to define the maximum distance for transporting wood on the basis of it being a percentage of the natural gas solution. We obtain:

$$CO_2^{ib}\left(D_{CO_2}^{H2020}\right) = 80\%\ CO_2^{ign} \tag{2.31}$$

The percentage 80% corresponds to a 20% reduction in CO_2 emissions in terms of iso-needs. We may do the same for emergy:

$$Em^{ws}\left(D_{Em}^{H2020}\right) = 80\%\ Em^{ngs} \tag{2.32}$$

If this reduction of 20% is exceeded, the technical solution will be better, according to the point of view considered, i.e. CO_2 emissions or emergy.

The relationship $\frac{D_{Em}^{H2020}}{D_{CO_2}^{H2020}}$ may be expressed as:

$$\frac{D_{Em}^{H2020}}{D_{CO_2}^{H2020}} = K^{H2020}\frac{\eta^w - \eta_{Em}^{min\,H2020}}{\eta^w - \eta_{CO_2}^{min\,H2020}} \tag{2.33}$$

with

$$K^{H2020} = \frac{\left(DEF^d + UFEF^d\right)\varepsilon\frac{M_{CO_2}}{M_C}}{LHV^d\,UEV^d}\frac{80\%Em^{ngs} - Em^{tr}(e) - Em(e) + Em^w(inv)}{80\%CO_2^{ngs} - CO_2^{tr}(e)} \tag{2.34}$$

$$\eta_{Em}^{min\,H2020} = Q_{th}\frac{\left(UEV^w + \frac{\tau}{PCI^w C^{ash}}D^{ash}\left(1+\gamma^{ash}\right)C_S^{ash}LHV^d\,UEV^d\right)}{80\%Em^{ngs} - Em^{tr}(e) - Em(e) - Em^w(inv)} \tag{2.35}$$

$$\eta_{CO_2}^{min\,H2020} = Q_{th}\frac{\left(\frac{\tau}{LHV^w C^{ash}}D^{ash}\left(1+\gamma^{ash}\right)C^{ash}\left(DEF^d + UFEF^d\right)\varepsilon\frac{M_{CO_2}}{M_C} + UFEF^w\right)}{80\%CO_2^{ing} - CO_2^{tr}(e)} \tag{2.36}$$

2.1.4. *Case study*

In Nantes, five establishments on the "La Chantrerie" site were structured so as to study the relevance of a wood fuel heating system. These establishments produce their main needs from natural gas on a site-by-site basis. This involves comparing a wood fuel system with a heating network to a decentralized gas-fired boiler, as shown in Figure 2.2.

Figure 2.2. *Example of a decentralized heating plant*

As the operations are in existence, data analysis for the three previous years provides a reference point Q_{th}. Necessary data collected for the emergy evaluation and the CO_2 assessment is given in Table 2.1.

1	Annual heating demands	4.28E+07	MJ		(*)
2	Annual heat exergy	8.25E+06	MJ		(*)
3	Humidity	25–45	%		
4	Molar mass relationship CO_2/C	44/12	g/g		
5	LHV of wood	9.3-13.6	MJ/kg		
6	LHV of dry wood	19	MJ/kg		EPA [EPA 05]

7	Ash content	2	%	(*)
8	Wood boiler output	50–75	%	(*)
9	Natural gas boiler output	82	%	(*)
10	Diesel LHV	36.5	MJ/l	Yao *et al.* [YAO 05]
11	Diesel oxidation rate	99	%	EPA [EPA 05]
12	Wood-carrying lorry tonnage	18–50	T	Shunping *et al.* [SHU 10]
13	Ash-carrying lorry tonnage	7	T	Shunping *et al.* [SHU 10]
14	Annual distance between home/plant	13,200	km	(*)
15	Ash transport distance removed	50	km	(*)
16	Specific car fuel consumption	0.092	l/km	Shunping *et al.* [SHU 10]
17	Specific (wood) lorry consumption	0.168-0.318	l/km	Shunping *et al.* [SHU 10]
18	Specific (ash) lorry consumption	0.242	l/km	Shunping *et al.* [SHU 10]
19	No-load consumption factor (wood)	0.75	–	(*)
20	No-load consumption factor (ash)	0.75	–	(*)
21	Number of hours worked per employee	5,280	hours	(*)
22	Wood heating plant investment	3.5E+06/20	€	Lifespan 20 years
23	Natural gas boiler investment	1.23E+06/20	€	Lifespan 20 years

Data with a label (*) are obtained from three years of data.

Table 2.1. *Data comparing wood-fired heating plant and natural gas boilers*

CO$_2$ emission factors are available from a document published by ADEME (French Agency of Energy and Environment) [ADE 10] (see Table 2.2). The unit emergy values stem from literature; see Table 2.3.

1	Upstream CO$_2$ emissions factor (natural gas)	0.01	kg CO$_2$/MJ	ADEME [ADE 10]
2	Upstream CO$_2$ emissions factor (natural gas)	0.05	kg CO$_2$/MJ	ADEME [ADE 10]
3	Upstream C emissions factor (diesel)	0.08	kg C/l	ADEME [ADE 10]
4	Local C emissions factor (diesel)	0.73	kg C/l	ADEME [ADE 10]
5	Upstream CO$_2$ emissions factor (wood)	0.0036	kg CO$_2$/MJ	ADEME [ADE 10]

Table 2.2. *Emissions factor*

		Unit	Unit emergy value (seJ/unit)	Reference
R1	Wood	J	5.62E+04	Odum [ODU 96]
NR1	Natural gas	J	7.73E +04	Odum [ODU 96]
F1	Transport of natural gas	J	1.74E +04	Romitelli [ROM 99]
F2	Diesel	J	1.07E +05	Odum *et al.* [ODU 00]
F3	Work done	h	8.58E +13	Jamali-Zghal [JAM 13]
F4	Euro	€	1.20E+12	Andric *et al.* [AND 15]

Table 2.3. *Unit emergy value*

In Table 2.4, annual energy natural gas consumption within the five establishments is given (see equation [2.16]). CO$_2$ emissions, both upstream and local, are deducted (see equation [2.19]). By applying equations [2.17] and [2.18], it is possible to calculate the emergies which are generated.

NATUREL GAS: annual assessment		
Annual consumption of natural gas	5.22E+13	J
Upstream CO_2 emissions	5.22E+05	kg CO_2
Local CO_2 emissions	2.61E+06	kg CO_2
Annual total CO_2	3.13E+06	kg CO_2
Upstream emergy	9.08E+17	seJ/year
Natural gas emergy	4.03E+18	seJ/year
Annual emergy investment	7.38E+16	seJ/year
Total annual emergy	5.02E+18	seJ/year

Table 2.4. CO_2 and emergy assessment for natural gas boilers

WOOD: annual assessment		
LHV damp wood	12.34	MJ/kg
Mass of wood	5.34E+06	kg
Number of return trips	356	
Mass of ash	1.07E+05	kg
Number of return trips (ash)	15	
Emissions CO_2 (ash)	949	kg CO_2
CO_2 emissions (workers)	3,571	kg CO_2
Upstream CO_2 emissions (wood)	237,046	kg CO_2
CO_2 emissions during transportation (wood)		
50 km	29,100	kg CO_2
Max D (CO_2)	**4,966**	Km
Emergy (ash)	1.25E+15	seJ/year
Emergy transport (worker)	4.70E+15	seJ/year
Emergy wood	3.70E+18	seJ/year
Emergy employed	4.53E+17	seJ/year
Annual emergy investment	2.10E+17	seJ/year
Emergy transport (wood)		
50 km	3.83E+16	seJ/year
Max D (emergy)	**846**	Km
(*) lorry carrying capacity (wood): 15 T		

Table 2.5. CO_2 assessment and emergy for a wood-fueled boiler

In Table 2.5, we find that the low heating value of wood is retained (humidity was deemed to be 35%) by applying equation [2.4]. Wood mass is obtained by applying equation [2.3]. The return journey is evaluated by deeming the haulage payload is 15 tons. For transporting ash, the payload chosen is 7 tons. The various CO_2 emissions are obtained using equations [2.12]–[2.15].

Equation [2.25] makes it possible to calculate $D_{CO_2}^{max}$. The numerical application gives a value of 4,966 km. This distance shows that the iso-emissions of CO_2, a wood central heating system, are an effective solution. Equation [2.21] makes it possible to calculate D_{Em}^{max}. The numerical calculation gives a value of 846 km. This distance shows that, with iso-emergy, a wood central heating system is a practical solution. On the other hand, distance, which is based on emergy analysis, is nearly six times lower than that for CO_2 emissions. This factor which is even more remarkable, if taking the "energy-climate" package, the distance $D_{CO_2}^{H2020}$ (see equation [2.31]) is 3,890 km, and the distance D_{Em}^{H2020} (see equation) [2.32]) is −465 km. The negative value means that it is impossible to satisfy heating demands in reducing by 20% the emergy of a natural gas heating system. The unique solution to lower the baseline emergy generated to satisfy comfort needs, would be the improvement of buildings (and therefore the lowering of iso-comfort needs).

So as to compare analyses, it is possible to interpret a diagram $(\Delta Em, \Delta CO_2)$, with $\Delta CO_2 = CO_2^{ngs} - CO_2^{ws}(D)$ and $\Delta Em = Em^{ngs} - Em^{ws}(D)$.

In Figure 2.3, the distance to transport wood increases. The point (6.47E+17, 2.89E+06) corresponds to a "theoretical" situation in which the wood heating would be in the middle of the logging operation, and the customers very close to the operation. The more the transport distance increases, the more the relevance of the substitution of natural gas by wood diminishes.

It may also be interesting to calculate the outputs of heating plants, as defined by equations [2.29] and [2.30] for both iso-emergy and CO_2 iso-emissions.

So as to obtain the previous distances (see Table 2.1), the annual average output of natural gas boilers has been selected as 82% (an average obtained by taking a three-year period) and the output of a wood fueled boiler has been estimated as 65%. The output of a wood fueled boiler is subject to variation; humidity and structure may vary. Figure 2.4 has outlined the proportion $\dfrac{\eta^w - \eta_{Em}^{min}}{\eta^w - \eta_{CO_2}^{min}}$. The numerical calculation gives: $\eta_{Em}^{min} = 55.3\%$ and

$\eta_{CO_2}^{min} = 4.93\%$. In the specific example, if the output of the wood fueled boiler becomes lower than 55%, the substitution of a natural gas plant by a wood heating plant increases emergy. At this stage, it is necessary to distinguish the emergy quantity, as explained in section 1.5; the renewable share, the non-renewable share and the part coming from the economic system.

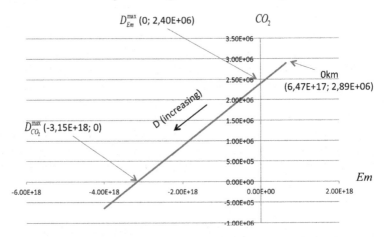

Figure 2.3. *Graph showing the inversely proportional relationship between CO_2 and the emergy of a wood heating system*

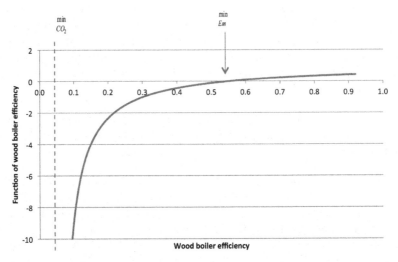

Figure 2.4. *Relationship between emergy distances and CO_2 (see equation [2.33])*

2.2. Wind resources

In the climate and energy package, one of the objectives is to develop renewable energies. Wind energy has the property of affording the strongest surface density in relation to the ground, the wind regime is favorable. In the following section, an emergy analysis of two wind turbines (of 850 kW and 3,000 kW) is set out.

2.2.1. *Emergy description*

The emergy diagram of a wind turbine is shown in Figure 2.5. The three stages in the life of a wind turbine are considered: its construction, its operation and its decommissioning. The total resources used at each of these stages correspond to the emergy of the electricity produced by the wind turbine. In what follows, the operating life of the wind turbine is considered over a 20-year period. The original work was reported by Paudel [PAU 14].

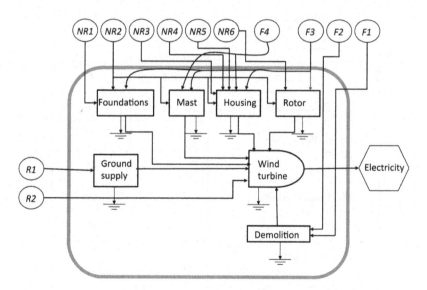

Figure 2.5. *Emergy diagram of a wind turbine*

The non-renewable resources are: concrete (*NR1*), steel (*NR2*), copper (*NR3*), aluminum (*NR4*), plastic (*NR5*) and fiberglass (*NR6*). The resources which come from the economic system are: services (*F1*), work (*F2*), fuel (*F3*) and paintwork (*F4*). Renewable resources are land possession (*R1*) and wind (*R2*).

Based on the work of Crawford [CRA 09], Paudel [PAU 14] carried out the inventory of resources used for a wind turbine, according to the nominal output, for an emergy analysis (see Tables 2.6 and 2.7).

So as to quantify the electricity production, there are two features to distinguish:

– an electricity production model;

– a wind resource (shown in the last row of Table 2.7).

		V52/850	V90/3,000
Nominal output	kW	850	3,000
Rotor diameter	M	52	90
Starting speed	m/s	3.5	4
Nominal speed	m/s	13	16
Maximum speed	m/s	25	25
Minimum mast size	m	59	65
Maximum mast size	m	78	105
Particular mast size	m	60	90
Number of blades		–	3
Length of blades	m	25	44

Table 2.6. *Features of wind turbines*

	Material		Unit	Unit emergy value	850 kW Quantity	850 kW Emergy seJ/year	3,000 kW Quantity	3,000 kW Emergy seJ/year
1	Concrete	NR1	G	2.41E+09	4.80E+08	1.16E+18	1.14E+09	2.75E+18
2	Steel	NR2	G	5.60E+09	1.09E+08	6.10E+17	2.76E+08	1.55E+18
3	Copper	NR3	G	1.14E+11	1.03E+06	1.17E+17	1.67E+06	1.90E+17
4	Aluminum	NR4	G	2.13E+10	5.99E+05	1.28E+16	2.01E+07	4.28E+17
5	Plastic	NR5	G	6.37E+08	2.19E+06	1.40E+15	8.73E+06	5.56E+15
6	Fiberglass and composites	NR6	G	1.32E+10	3.01E+06	3.97E+16	1.20E+07	1.58E+17
7	Services	F1	€	1.34E+12	9.25E+05	1.24E+18	3.26E+06	4.37E+18
8	Work done	F2	man-year	4.70E+17	2.11E+00	9.92E+17	7.46E+00	3.51E+18
9	Diesel	F3	G	4.95E+09	4.94E+05	2.45E+15	1.74E+06	8.61E+15
10	Paintwork	F4	G	2.52E+10	9.30E+05	2.34E+16	1.24E+06	3.12E+16
11	Land supply	R1	m²/year	1.34E+11	2,124	2.85E+14	6.36E+03	8.53E+14
12	Wind	R2	J	4.19E+03				

Table 2.7. Annual emergy for relevant wind turbine resources

2.2.2. Simplified model wind turbine

The driving force of a wind turbine is, of course, wind speed. The latter varies according to height. A power law model was put forward by Hellman:

$$V = V_{ref} \left(\frac{z}{z_{ref}} \right)^H \tag{2.37}$$

The baseline speed corresponds to the associated baseline height. For example, meteorological stations are located at 5 m, the rotors of the wind turbines in Table 2.6 are located respectively at 60 and 90 m, the measurement carried out at 5 m should therefore be adjusted. The Hellman coefficient considered to be taken as 0.28, as per the value in the literature [OZG 07]. Those constructing wind turbines give the feature between electric power W_e and the upstream wind speed V_1 [PED 12] (see Figure 2.6). This feature breaks down into four spheres. Before reaching the so-called starting speed, the wind turbine is idle (see Table 2.6). Between the starting speed and the speed which produces the nominal output of the wind turbine, production increases in an approximately linear way with the wind speed (upwind). Between the nominal speed and the maximum speed (which enables the protection of the wind turbine in case of storms, for example), the wind turbine production is managed so as to maintain the nominal value.

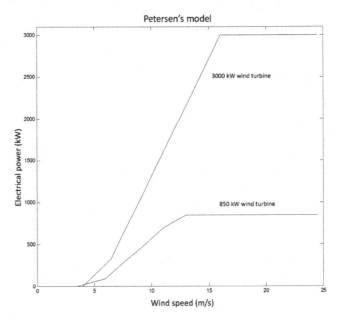

Figure 2.6. *Electrical power according to wind speed. For a color version of the figure, see www.iste.co.uk/lecorre/emergy.zip*

The electrical power produced W_{el} by the wind turbine is equal to the of the air flow driving the blades $\dot{m}_{air}$ multiplied by the difference in speed between the upwind speed V_1 and the downwind speed V_2, see Figure 2.7 for the notations, and the average speed $\overline{V} = \dfrac{V_1 + V_2}{2}$. We obtain resultant power by:

$$W_{el} = \dot{m}_{air}\left(V_1 - V_2\right)\overline{V} \qquad [2.38]$$

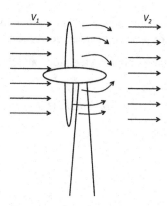

Figure 2.7. *Simplified model of a wind turbine*

The airflow driving the blades $\dot{m}_{air}$ is equal to:

$$\dot{m}_{air} = \rho_{air} \; SB \; \bar{V} \qquad\qquad [2.39]$$

with ρ_{air} the air density, SB the surface swept by the blades, i.e. $SB = \frac{\pi \, l_b^2}{4}$, with l_b the length of the blades. By introducing the velocity relationship Rv which is stated as $Rv = \frac{V_1}{V_2}$, the upwind speed is derived from the equation solution:

$$(1 + Rv)^2 \, (1 - Rv) = 4\frac{W_{el}}{\rho_{air} \; SB \; V_1^3} \qquad\qquad [2.40]$$

This equation was devised by Betz [BET 46] and makes it possible to demonstrate that the maximum power for a wind turbine is obtained by a downwind speed which is 1/3 of the upwind speed.

In Figure 2.8, the difference in speed is outlined according to the upwind speed through specific calculations [PED 92] on the one hand, and of the third-order polynome within equation [2.40], we obtain three spheres:

– as long as the upwind speed is lower than the starting speed, there is no difference in speed;

– between the starting speed and the nominal speed; the difference in speed increases, up to the maximum speed (10 m/s in Figure 2.8) then decreases;

– so as to produce the same power for increasing upwind speeds (between 16 m/s and 24 m/s), the difference in speed decreases;

– the wind turbine is then in protected mode and the difference in speed is again nil.

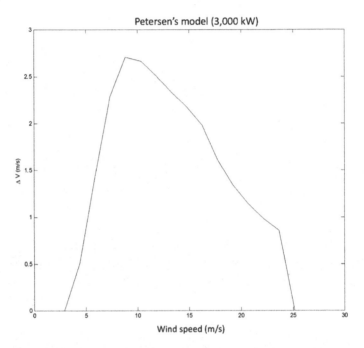

Figure 2.8. *Difference in upwind-downwind turbine speed according wind speed*

Figure 2.9 shows the electricity production of the 3,000 kW wind turbine of according to the difference in speed. This feature demonstrates the previous spheres very well.

2.2.3. *Wind resource*

Wind resource depends on location. This information is available in the database on the website of the US Department of State for Energy. In the following sections, the results for 26 European countries, representing 90 meteorological stations, have been shown (see the map of weather stations in Figure 2.10). The average annual wind speeds at 5 m are given in Figure 2.11.

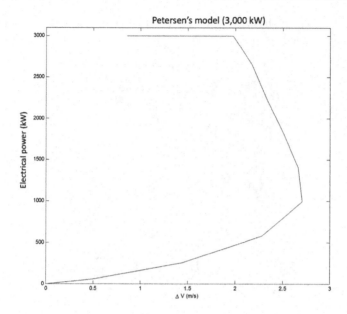

Figure 2.9. *Electricity production according to the difference in upwind and downwind wind speeds*

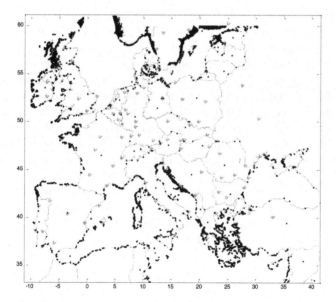

Figure 2.10. *Location of relevant meteorological stations*

Figure 2.11 emphasizes all of the coasts of Great Britain, but also the mistral in the south of France and Portugal.

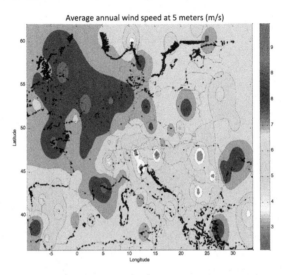

Figure 2.11. *Average annual wind speed across Europe. For a color version of the figure, see www.iste.co.uk/lecorre/emergy.zip*

Having wind resources available (upwind at a height of 5 m), it is possible to calculate electricity production from Pedersen *et al.*'s model [PED 92]. This electricity production corresponds to the output in the emergy diagram in Figure 2.5. It is thus possible to complete the latter for each meteorological station. The electrical transformity produced by a wind turbine depends on its height (see Table 2.7) and its location. It is possible to establish an average annual transformity across Europe by turbine range (see Figures 2.12 (850 kW wind turbine) and 2.13 (3,000 kW wind turbine)). The lowest transformities are achieved by those regions with the strongest winds. Whatever the location there will always be an "emergy" investment before the first kilowatt is produced. It then follows that the more favorable the site is to wind, the greater the production significance and the weaker the transformity of electricity. Likewise, the transformity of electricity of a wind turbine is generally greater than that of a wind turbine of 3,000 kW (however, a given wind condition may lead to qualification of this assertion).

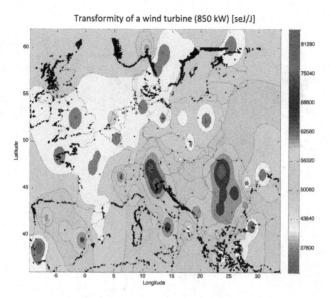

Figure 2.12. *Electricity transformity produced by an 850 kW wind turbine. For a color version of the figure, see www.iste.co.uk/lecorre/emergy.zip*

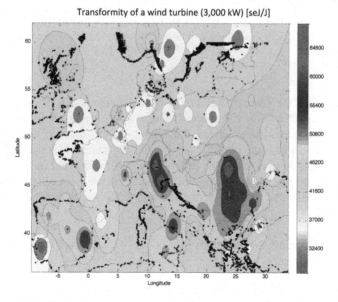

Figure 2.13. *Electricity transformity produced by a 3,000 kW wind turbine. For a color version of the figure, see www.iste.co.uk/lecorre/emergy.zip*

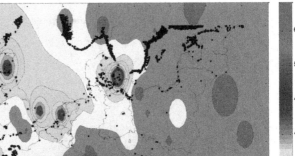

Figure 2.14. *Emergy Sustainability Index (ESI) for electricity produced by an 850 kW wind turbine. For a color version of the figure, see www.iste.co.uk/lecorre/emergy.zip*

The greater the value of the Emergy Sustainability Index (ESI), defined in Chapter 1, section 1.6, the more the local resource (even the renewable resource) is generated to create the output (to be specific, in this case, electricity). In Figures 2.14 and 2.15, the ESI is outlined. It is noticeable that for this index, there is almost no difference in area with wind turbines of 850 and 3,000 kW, but there remains a difference in value (between 0 and 6 for a wind turbine of 850 kW and between 0 and 8 for that of a wind turbine of 3,000 kW).

By introducing the notations which are brought in from section 1.5, Table 2.7 gives us:

– goods and services: S={items 1 to 10};

– renewable resource: R={item 11 and 12}.

Emergy Sustainability Index ESI(3,000 kW)

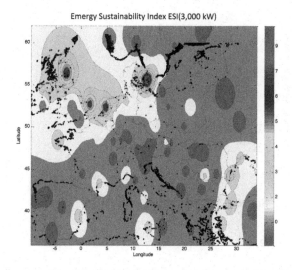

Figure 2.15. *Emergy Sustainability Index (ESI) for electricity produced by a 3,000 kW wind turbine. For a color version of the figure, see www.iste.co.uk/lecorre/emergy.zip*

For Brest, in Brittany (France), we find the following ratios.

	Wind turbine	
	850 kW	3,000 kW
$EYR=(F+R)/F$	2.38E+00	2.69E+00
$ELR=F/R$	7.26E-01	5.93E-01
ESI	3.28E+00	4.53E+00

Table 2.8. *Emergy ratios for wind turbines on the Brest site*

In Figure 2.16, the Emergy Sustainability Index is shown according wind turbine power. It grows with the height of the wind turbine.

To conclude this section, the analysis proposed on a continental scale shows, as the main pitfall, the initial grid pattern of weather stations. The emergy approach can apply to various spatial scales over the lifetime of a system. The overall results presented must be confirmed by studying a specific site with a specific and localized database. Nevertheless, the results shown make it possible to identify the most relevant sites from an emergy point of view.

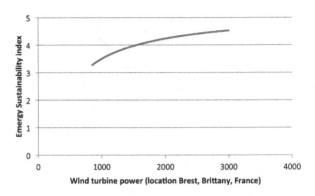

Figure 2.16. *Sustainability index for the Brest site, wind turbine 3,000 kW*

2.3. Solar panels

Paoli *et al.* [PAO 08] carried out the emergy evaluation of a thermal solar power station and a photovoltaic power station.

2.3.1. *Thermal solar power station*

The power station studied is made up of 65 solar panels over a surface area of 136 m^2 with a storage reservoir of 4,000 liters (see Figure 2.17). The lifetime of the plant is estimated to be 20 years. Each solar panel is made up of eight emergy sources the main ones of which are copper, glass and aluminum as well as the work element.

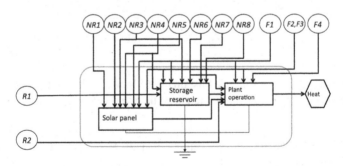

Figure 2.17. *Emergy diagram thermal solar panel[1]*

1 Notations are included in Tables 2.9 to 2.11.

In Table 2.9, the "quantity" column corresponds to the panel unit quantity, the "emergy" column to the quantity output by unit "emergy" value, the "annual emergy" column to the "emergy" column divided over the power station's lifetime. The last column is the annual emergy for all 65 panels.

Item		Quantity	Unit	UEV	Emergy	Annual emergy	Power station
Panel	Component						
NR1	Glass	2.02E+04	g	1.90E+09	3.84E+13	1.92E+12	1.25E+14
NR2	Aluminum	7.01E+03	g	1.27E+10	8.90E+13	4.45E+12	2.89E+14
NR3	Copper	6.10E+03	g	6.77E+10	4.13E+14	2.06E+13	1.34E+15
NR4	Polyurethane	2.44E+03	g	5.87E+09	1.43E+13	7.16E+11	4.65E+13
NR5	Mineral wool	2.16E+03	g	5.73E+09	1.24E+13	6.19E+11	4.02E+13
F1	Electricity	6.57E+07	J	1.74E+05	1.14E+13	5.72E+11	3.72E+13
F2	Fuel	2.67E+08	J	6.60E+04	1.76E+13	8.81E+11	5.73E+13
F3	Work done	1.40E+07	J	7.38E+06	1.03E+14	5.17E+12	3.36E+14
					Total emergy (seJ/y)		2.27E+15

Table 2.9. *Thermal assessment of a thermal solar power station: [PAO 08]*

The thermal power station is equipped with a substantial energy storage reservoir, with a volume of 4,000 liters. This reservoir is made up of six emergy sources, the main ones of which are steel, copper and electricity (see Table 2.10). The construction cannot be distinguished from that of the plant and is incorporated within Table 2.11.

Item		Quantity	Unit	UEV	Emergy	Annual emergy
Storage	Component					
NR6	Steel	4.91E+05	G	4.15E+09	2.04E+15	1.02E+14
NR4	Polyurethane	3.32E+04	G	5.87E+09	1.95E+14	9.74E+12
NR7	Propylene Glycol	1.27E+04	G	3.80E+08	4.83E+12	2.41E+11
NR8	Copper	2.63E+04	G	6.77E+10	1.78E+15	8.90E+13
R1	Water	2.46E+04	G	7.30E+06	1.80E+11	8.98E+09
F1	Electricity	2.83E+09	J	1.74E+05	4.92E+14	2.46E+13
				Total emergy (seJ/y)		2.26E+14

Table 2.10. *Emergy assessment significant storage reservoir*

The assembly, installation and operation of the power station requires five emergy sources, work and other related costs there being two main ones (see Table 2.11).

Item		Quantity	Unit	UEV	Emergy	Annual emergy
m2	m2	Seedling				
NR6	Steel	1.29E+04	g	4.15E+09	5.35E+13	2.68E+12
F1	Electricity	4.77E+06	J	1.74E+05	8.30E+11	4.15E+10
F2	Diesel	6.33E+07	J	6.60E+04	4.18E+12	2.09E+11
F3	Work done	1.44E+07	J	7.38E+06	1.06E+14	5.31E+12
F4	Other costs	2.45E+01	€	2.22E+12	5.44E+13	2.72E+12
				Total emergy (seJ/y)		1.10E+13

Table 2.11. *Emergy assessment of installation and operation of the power station [PAO 08]*

Authors in the field [PAO 08] located the power station in the Liguria region in north Italy. The supply of sunlight is given in Table 2.12.

Item		Quantity	Unit	UEV	Emergy	Annual emergy	Power station
R^2	Sunlight	3.29E+11	J	1	3.29E+11	2.19E+10	1.43E+12

Table 2.12. *Solar power station provision [PAO 08]*

The overall annual emergy is the sum of all sources, with the value of 2.51E+15 seJ/year. The estimated production is 1.78E+11 J/year. Assuming that the hot water produced is 45°C and that the baseline temperature is 15°C, we obtain a CARNOT factor of 9.44%. The annual exergy is 1.68E+10 J/year and the heat transformity for this power station has the value of 1.50E+05 seJ/J.

2.3.2. Photovoltaic power station

The photovoltaic power station has the same surface area as described previously (136 m^2). Figure 2.18 shows the emergy diagram associated with the analysis.

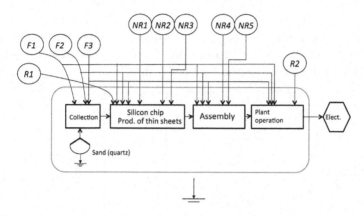

Figure 2.18. *Emergy diagram of a photovoltaic cell*

Without any loss of generality, the sources are grouped together within the same classification for the sake of clarity in Figure 2.18. The components of the so-called wafer are given in the emergy table (see Table 2.13).

Item Wafers		Panel	Unit	UEV (seJ/unit)	Emergy seJ	Annual emergy seJ/year
NR	Quartz/silica (sand)	2.84E+03	g	1.00E+09	2.84E+12	1.89E+11
NR3	Coke	1.11E+07	J	4.00E+04	4.44E+11	2.96E+10
NR3	Charcoal	2.41E+07	J	1.06E+05	2.55E+12	1.70E+11
NR3	Graphite	1.30E+02	g	3.15E+09	4.10E+11	2.73E+10
NR3	Wood	1.32E+03	g	8.79E+08	1.16E+12	7.74E+10
NR2	Polyethylene	6.36E+01	g	5.87E+09	3.73E+11	2.49E+10
NR1	HCl	5.99E+02	g	3.64E+09	2.18E+12	1.45E+11
R1	Water	1.40E+03	g	7.30E+06	1.02E+10	6.81E+08
NR1	NaOH	5.80E+00	g	1.90E+09	1.10E+10	7.35E+08
NR1	H2SO4	4.30E+00	g	3.64E+09	1.57E+10	1.04E+09
NR1	POCl3	6.00E+02	g	1.01E+09	6.06E+11	4.04E+10
NR1	HF	1.10E+00	g	9.89E+08	1.09E+09	7.25E+07
NR1	CF4	7.00E+02	g	1.01E+09	7.07E+11	4.71E+10
NR1	Ag/Al glue	6.00E+01	g	1.69E+10	1.01E+12	6.76E+10
F3	Natural gas	1.22E+08	J	4.80E+04	5.86E+12	3.90E+11
F3	Electricity	7.49E+08	J	1.74E+05	1.30E+14	8.69E+12

Table 2.13. *Emergy assessment of a wafer [PAO 08]*

The components of a photovoltaic panel are given in the emergy table (see Table 2.14). There are currently a total of 215 photovoltaic panels. The lifetime of these panels is estimated to be 15 years. The calculation is similar to that presented for the thermal solar power station.

Item	Cell assembly	Panel	Unit	UEV (seJ/unit)	Emergy	Annual emergy
NR5	Aluminum	2.01E+03	g	1.27E+10	2.55E+13	1.70E+12
F2	Glass	4.05E+03	g	1.90E+09	7.70E+12	5.13E+11
NR4	Ethylene-vinyl acetate	5.38E+02	g	5.87E+09	3.16E+12	2.11E+11
NR4	Tedlar (Polyvinyl fluoride)	6.14E+01	g	6.32E+09	3.88E+11	2.59E+10
F2	Steel	1.57E+04	g	4.15E+09	6.52E+13	4.34E+12
F2	Copperplate	2.52E+01	g	9.90E+10	2.49E+12	1.66E+11
NR4	Plastic	2.52E+01	g	5.87E+09	1.48E+11	9.86E+09
F1	Work done	1.36E+06	J	7.38E+06	1.00E+13	6.69E+11

Table 2.14. *Emergy assessment of a photovoltaic panel [PAO 08]*

The emergy Table 2.15 makes it possible to evaluate the annual emergy for the construction and operation of the power station.

Item	Assembly/operation of the power station	Panel	Unit	UEV (seJ/unit)	Emergy	Annual emergy
F3	Fuel	6.80E+06	J	6.60E+04	4.49E+11	2.99E+10
F2	Reverser	4.77E+01	€	2.22E+12	1.06E+14	7.06E+12
F1	Maintenance costs	4.18E+01	€	2.22E+12	9.28E+13	6.19E+12
F1	Work done	5.40E+05	J	7.38E+06	3.99E+12	2.66E+11
F1	Feasibility study	1.08E+00	€	2.22E+12	2.40E+12	1.60E+11

Table 2.15. *Assembly and operation of a photovoltaic power station*

Solar resources are shown in emergy Table 2.16.

Item		Power station	Unit	UEV (seJ/unit)	Emergy	Annual emergy
R^2	Sunlight	6.58E+10	J	5.62E+10	6.58E+10	3.29E+09

Table 2.16. *Solar power station provision [PAO 08]*

The annual emergy of the power station is 6.72E+15seJ/year for electricity generation of 5.62E+10J. The transformity of this power station is therefore 1.20E+05 seJ/J. We reach the counter-intuitive result of the electricity transformity produced by a photovoltaic power station being lower than that of a thermal solar power station. Regarding this emergy point of view, electricity production rather than heat production is preferable, when considering transformity as a criteria. This conclusion might be qualified by the analysis of ratios which are introduced in section 1.6.

2.4. The production of biodiesel and glycerol from palm oil

Plants are a means of converting and storing solar energy. The production of so-called first generation fuel involved resorting to usable plants in the food industry. Nimmanterdwong *et al.* [NIM 15] assessed the production of fuel from palm oil and Jatropha[2].

In what follows, we only detail the emergy analysis carried out for palm oil. The reader is invited to refer to the article on Jatrophas: the methodology is entirely the same.

The first "conversion" corresponds to plant development. This part integrates renewable resources (sun and rain), seeds, fertilizers and herbicides and fossil energy for agricultural production. After being harvested, the palm is transported to the processing plant as oil. Part of the oil produced may then be intended for consumption (35%), or for conversion into biodiesel. The glycerol obtained then becomes a co-product of this stage. Next, all of the products are transported to their end use. Figure 2.19 gives an emergy representation of biodiesel production from resources generated during the agricultural phase.

---

2 The majority of these species of plants are toxic but have medicinal qualities.

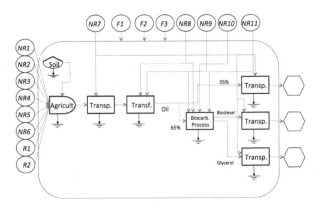

Figure 2.19. *Emergy diagram for the production
of bio-alcohol from palm oil*

All of the constituents of a bio-alcohol production plant are shown in the emergy table (see Table 2.17). Services (financial flows) are subdivided into three major families (work *F1*, various services *F2* and equipment *F3*) over three stages: agricultural activity, conversion into oil and the process of producing bio-fuels. Unit emergy values associated with these flows are of a given currency and are therefore identical.

		Quantity	Unit	UEV	seJ/unit	Emergy seJ/hectare/year
R1	Sunlight	6.64E+13	J/hectare/year	1	seJ/J	6.64E+13
R2	Rain	7.37E+10	J/hectare/year	3.06E+04	seJ/J	2.26E+15
–	Soil	5.96E+09	J/hectare/year	2.25E+05	seJ/J	1.34E+15
NR1	Seed	168	kg/hectare/year	2.67E+11	seJ/kg	4.49E+13
NR2	Nitrogen	119.85	kg/hectare/year	6.38E+12	seJ/kg	7.65E+14
NR3	Phosphate	0.8	kg/hectare/year	6.55E+12	seJ/kg	5.24E+12
NR4	Potash	60.26	kg/hectare/year	1.85E+12	seJ/kg	1.11E+14
NR5	Herbicide	4.76	kg/hectare/year	2.48E+13	seJ/kg	1.18E+14
NR6	Pesticide	1.7	kg/hectare/year	2.48E+13	seJ/kg	4.216E+13
NR8	Water for irrigation	1.46E+01	kg/hectare/year	2.03E+08	seJ/kg	2.96E+09
	Steam	5.54E+09	J/hectare/year	1.69E+05	seJ/J	9.36E+14
	Water (cooling)	1.25E+04	kg/hectare/year	6.64E+08	seJ/J	8.30E+12
NR9	MeOH	3.04E+02	kg/hectare/year	7.23E+12	seJ/kg	2.20E+15
NR10	Soda NaOH	3.13E+00	J/hectare/year	6.38E+12	seJ/kg	2.00E+13
NR11	Electricity	2.93E+08	J/hectare/year	1.69E+05	seJ/J	4.95E+13

NR7	Agricultural diesel	6.81E+09	J/hectare/year	1.10E+05	seJ/J	7.49E+14
	Conversion	1.12E+08	J/hectare/year	1.10E+05	seJ/J	1.23E+13
	Process	5.55E+09	J/hectare/year	1.10E+05	seJ/J	6.11E+14
F1	Agricultural work	341.57	$/hectare/year	1.49E+13	seJ/$	5.09E+15
	Work required for transport	30.201	$/hectare/year	1.49E+13	seJ/$	4.50E+14
	Processing work required	69.25	$/hectare/year	1.49E+13	seJ/$	1.03E+15
F2	Various agricultural services	145.35	$/hectare/year	1.49E+13	seJ/$	2.17E+15
	Conversion	50.35	$/hectare/year	1.49E+13	seJ/$	7.50E+14
	Process	123.32	$/hectare/year	1.49E+13	seJ/$	1.84E+15
F3	Agricultural machinery	24.22	$/hectare/year	1.49E+13	seJ/$	3.61E+14
	Conversion	45.74	$/hectare/year	1.49E+13	seJ/$	6.82E+14
	Process	95.81	$/hectare/year	1.49E+13	seJ/$	1.43E+15

Table 2.17. *Emergy assessment of the agricultural phase [NIM 15]*

				Emergy seJ/hectare/year
Oil	Quantity	3.03E+03	kg/hectare/year	
	LHV (MJ/kg)	40.14		1.50E+16
	Energy	1.21E+11	J/hectare/year	
		UEV (palm)	1.24E+05 seJ/J	
Co-product	Biodiesel	2.85E+03	kg/hectare/year	
	LHV (MJ/kg)	38.07		2.31E+16
	Energy	1.08E+11	J/hectare/year	
		UEV (biodiesel)	2.13E+05 seJ/J	
Co-product	Glycerol	512.8	kg/hectare/year	
	LHV (MJ/kg)	3.42		
	Energy	1.75E+09	J/hectare/year	
		UEV (Glycerol)	**1.32E+07 seJ/J**	

Table 2.18. *Unit emergy value of palm oil, biodiesel or glycerol [NIM 15]*

2.5. Production of biofuel from algae

After the controversy regarding first generation biofuels, algae were seen as a conversion driver away from solar energy. Macro-algae and micro-algae can be differentiated.

2.5.1. *Macro-algae*

Seghetta *et al.* [SEG 14] carried out an emergy analysis linked to bioethanol production of macro-algae. Two sites were studied, one in Denmark and the other in Italy. The Danish site is described below. The reader is invited to read the article on the Italian site. The main difference between the sites is that crates of algae necessitated artificial pumping on the Italian site. Figure 2.20 shows the emergy description for natural production of algae and the associated conversion.

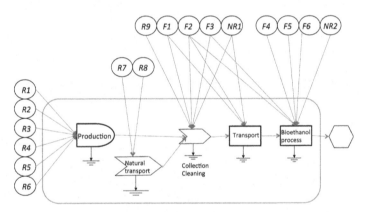

Figure 2.20. *Natural algae production and associated conversion*

The production of macro-algae is natural: nitrogen and phosphorous are considered to be renewable resources, unlike the previous study, in section 2.4, in which humans supplied the fertilizer and pesticides used during palm tree production. Once produced, the algae were washed and dried and then transported to the bioethanol production plant.

The emergy Table 2.19 allows us to make a number of comments:

– wind is considered within the inventory of sources. If we refer to the geobiosphere model in Chapter 1 (indeed according to Brown and Ulgiati's original model [BRO 10]), this term might be a co-product the primary sources and should appear having a maximum. Nevertheless, the *R3* source results from a global assessment of the geobiosphere. It is possible to consider sources *R1-R4* as genuinely independent;

– sources *R7* and *R8*, the waves and tides, when considered in the same perspective as previously, are seen as both independent (and independent of sources *R1–R4*);

– sources *F1–F3* are considered as coming from the economic system although the authors have not used the currency to distinguish these flows. These sources as well as the source *NR1* being resorted to at each stage, the respective contributions are explained.

The unit energy value is calculated upon knowing the annual production of bioethanol (see Table 2.20).

		Quantity	Unit (year)	UEV seJ/unit	Emergy seJ/year
R1	Sunlight	3.09E+17	J	1	3.09E+17
R2	Rain	4.62E+13	g	4.80E+05	2.22E+19
R3	Wind	5.92E+14	J	9.83E+02	5.81E+17
R4	Geothermal science	1.47E+14	J	1.20E+04	1.76E+18
R5	Nitrogen	1.95E+08	g	7.73E+09	1.51E+18
R6	Phosphorous	1.04E+07	g	2.99E+10	3.12E+17
R7	Waves	5.11E+14	J	5.12E+04	2.62E+19
R8	Tides	2.66E+13	J	7.37E+04	1.96E+18
R9	Surface drying	6.97E+10	g	1.14E+06	7.95E+16
		5.13E+09	g	1.14E+06	5.85E+15
F1	Equipment	2.33E+05	g	9.22E+09	2.15E+15
		1.44E+03	g	9.22E+09	1.33E+13
		7.07E+04	g	9.22E+09	6.52E+14
F2	Work done	8.35E+02	h	6.50E+12	5.43E+15
		1.67E+03	h	6.50E+12	1.09E+16
		8.35E+02	h	6.50E+12	5.43E+15
		1.19E+04	h	6.50E+12	7.74E+16
F3	Electricity	2.01E+12	J	2.69E+05	5.41E+17
		4.24E+12	J	2.69E+05	1.14E+18
NR1	Diesel	1.42E+07	g	4.89E+09	6.94E+16
		3.41E+07	g	4.89E+09	1.67E+17
		3.92E+06	g	4.89E+09	1.92E+16
NR2	Natural gas	2.96E+13	J	6.72E+04	1.99E+18
F3	Steel	1.69E+08	g	4.89E+09	8.26E+17
F4	Concrete	3.00E+08	g	1.81E+09	5.43E+17
F5	Oil	7.28E+10	J	9.07E+04	6.60E+15

Table 2.19. *Emergy assessment of bioethanol production by macro-algae [SEG 14]*

					Emergy seJ/year
Bioethanol	Energy	2.32E+13	J		5.91E+19
		UEV (bioethanol)	**2.55E+06**	seJ/J	

Table 2.20. *Unit emergy value of bioethanol obtained from algae [SEG 14]*

2.5.2. *Micro-algae*

There is a very large variety of micro-algae (between 200,000 and 800,000 according to Delrue [DEL 13]) which yields a per hectare oil production of between 3 and 10 times greater than that obtained from palm trees. [ALV 12] put forward an emergy analysis for oil production so as to then make biodiesel from lipid micro-algae by using centrifuge technology (see Figure 2.21).

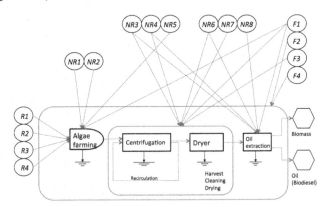

Figure 2.21. *Production of biodiesel from micro-algae*

The micro-algae culture, envisaged by Alves da Cruz and Oller do Nascimento [ALV 12], is the result of human activity. Nutrients necessary for the development of the latter (nitrogen and phosphorous) are therefore included as non-renewable resources (a notable difference compared with Seghetta *et al.*'s study [SEG 14]).

Micro-algae are not usually found to be concentrated in their environment. A centrifuge enables the latter to be concentrated with a preparation stage: harvesting, washing and drying. The separation technique in respect of the lipid portion is then carried out.

The emergy assessment of biodiesel production from micro-algae is synthesized in the emergy table (Table 2.21). The make-up of the emergy table is similar to the preceding study. It may be noted that work is not distinguished by currency but rather as an energy form.

		Quantity	Unit (year)	UEV seJ/unit	Emergy seJ/year
R1	Sunlight	2.59E+16	J	1	2.59E+16
R2	Rain	5.31E+09	kg	1.51E+08	8.02E+17
R3	Geothermal science	1.50E+13	J	1.01E+04	1.52E+17
R4	Water	6.73E+09	kg	1.25E+09	8.41E+18
NR1	Nitrogen	3.98E+06	kg	2.41E+10	9.59E+16
NR2	Phosphorous	3.11E+05	kg	1.13E+10	3.51E+15
NR3	Concrete	4.89E+05	kg	9.26E+10	4.53E+16
NR4	Scrap iron	7.14E+03	kg	1.13E+13	8.07E+16
NR5	Steel	6.24E+04	kg	1.13E+13	7.05E+17
NR6	Electricity	2.14E+14	J	2.00E+05	4.28E+19
NR7	Natural gas	5.66E+14	J	4.35E+04	2.46E+19
NR8	Hexane	2.06E+06	kg	6.08E+12	1.25E+19
F1	Work done	1.33E+10	J	1.24E+07	1.65E+17
F2	Construction	1.55E+06	$	1.10E+12	1.71E+18
F3	Service	1.64E+07	$	1.10E+12	1.80E+19
F4	Tax	3.11E+06	$	1.10E+12	3.42E+18

Table 2.21. *Emergy assessment of biodiesel production from micro-algae [ALV 12]*

			Emergy seJ/year
			1.14E+20
Co-products			
Biomass	3.16E+07		kg
	LHV	1.70E+01	MJ/kg
	UEV (seJ/year)		2.12E+05
Biodiesel	9.70E+06		kg
	LHV	3.35E+01	MJ/kg
	UEV (seJ/year)		**3.50E+05**

Table 2.22. *Unit emergy value of biodiesel from micro-algae [ALV 12]*

Table 2.22 makes it possible to obtain the unit emergy value of biodiesel from micro-algae.

2.6. Synthesis

All of the examples described in this chapter are synthesized in Table 2.23: the first column relates to the technology type, the second to the unit emergy value and the remaining columns the *EYR, ELR, ESI* and *%R ratios*. Photovoltaic technology emerges as the emergy resource with the lowest consumption, thereby being the most relevant from an emergy point of view. It is then possible to group together wind, the production of biofuels from palm oil or from micro-algae. The latter outs forward interesting ratios, particularly the ratio *%R*. Thermal solar energy is something of an anomaly. This technology's output is heat.

Table 2.23 elicits various comments:

– the unit emergy values for the emergy production from wind and photovoltaic technologies are transformities. Likewise, for thermal solar, a Carnot factor has been applied So as to obtain electricity from bio-resources or fossil fuels, the conversion factor must be taken account of;

– as expected, solar is becoming the resource using the least energy so as to obtain electricity/exergy. There is only a 20% difference between solar thermal and photovoltaic technology;

– the unit emergy value of sugar cane depends on the location of the crop in significant quantities, 4.9 to 23.5E+04 seJ/J;

– agricultural resources obtain a unit energy value on a par with those of fossil resources. Nevertheless second or third generation biofuels are co-products of cereals, or indeed of other products intended for use in food. These unit emergy values in the comprehensive emergy diagram would therefore then be included using the maximum capacity processing unit. Please see Chapter 1. This is absolutely not the case for fossil resources. Thus, it is not so much the absolute unit emergy value which must be kept in an analysis but indeed its production history: a fundamental element of emergy analysis; and

– the EYR production ratio for agro-fuels is close to a single unit: this means that intensive agriculture (the provision of fertilizers and pesticides) generates a huge amount of goods and services which come from the economic system (and barely falling within the notion of *sustainable development*). On the basis of analyzing this value, such agro-fuels might be implemented. This is only relevant when accounting for the entire system operation (by means of co-products).

	UEV (1E+04 seJ/J)	EYR	ELR	ESI	%R	Reference
Wind(*)	37.5–81.3	2.5	0.8	2–9	138	section 2.2
Thermal solar	15(**)	5.4	1,760	3.06	5.68E-02	section 2.3
Solar Photovoltaic	12	1.36	9.4	1.43E-04	1.05E-02	section 2.4
Palm	21.3	1.43	5.31	0.27	16	
Algae (macro)	255	20.9	0.14	153.2	88	section 2.5
Algae (micro) (***)	35	4.9	11.1	0.44	8	
Sugarcane (Brazil)	4.9	1.57	–	–	31	Zhang and Long [ZHA 10]
Sugarcane (Louisiana)	15.6	–	–	–	–	Bastianoni and Marchettini [BAS 96]
Sugarcane (Brazil)	17.3	–	–	–	–	Bastianoni and Marchettini [BAS 96]
Sugarcane (Florida)	23.5	–	–	–	–	Bastianoni and Marchettini [BAS 96]
Corn (China)	27.7	1.24	–	–	20	Zhang and Long [ZHA 10]
Maize (Italy)	18.9	1.14	–	–	11	Zhang and Long [ZHA 10]
Corn (Denmark)	10.7	–	–	–	–	Coppola *et al.* [COP 09]
Hop – flexible climbing plant stem (Italy)	128	–	–	–	–	Bastianoni and Marchettini [BAS 96]
Straw (Denmark)	14.7–50.7	–	–	–	–	Coppola *et al.* [COP 09]
Carbon coal	12	–	–	–	–	
Oil	18	–	–	–	–	see Chapter 1
Natural gas	17.8	–	–	–	–	

(*) According to the turbine power and site.
(**) Converted with a Carnot efficiency.
(***) Calculations were made by the author.

Table 2.23. *Overview of bio-energy, solar and wind resources*

In Table 2.22, the unit emergy value of fossil fuels was stated. Through the comparison between energy biomass, sugarcane, wheat, maize, hops (flexible climbing plant stem) or straw with fossil fuels, it is noted that first generation biofuels might have unit emergy values lower than those of fossil fuels. Nevertheless, biofuels and fossil fuels are on a par in terms of unit emergy value. This means that only so-called "efficient" energy savings in the emergy context should be sought. Substituting fuels is not opportune from an emergy point of view nor by the meaning of the Bruntland Commission report [WCE 87].

Emergy and Recycling

Recycling is a major issue in the effort to limit reliance on terrestrial resources. How can emergy analysis incorporate the history of a material, in which the residue, waste, or even products of an upstream system become the input materials for a downstream system? This chapter has been designed with graduated difficulty; that is, from *continuous* recycling with no loss to *discontinuous* recycling with loss. The meaning of *continuous* and *discontinuous* (batch) has to do with the material's process. In the case referred to as *continuous*, material may either be sent to a landfill site or be recycled, while in the *discontinuous* case, recycled material is stored, and materials that have undergone various different processes may mix.

3.1. Introduction

The BS EN ISO 14044 (2006) certification proposes a distinction between:

– *Open-loop* recycling: a product at the end of its life is converted into another product, generally of lower quality [SHE 10]. For example, the components of milk cartons are separated (polyethylene-aluminum composite and cardboard): the cardboard is converted into toilet paper and paper towels, while the polyethylene-aluminum composite is converted into garden furniture, etc.;

– *Closed-loop* recycling: a product at the end of its life is remade into another product of the same type [MEN 06].

On this basis, this chapter will focus in particular on closed-loop recycling.

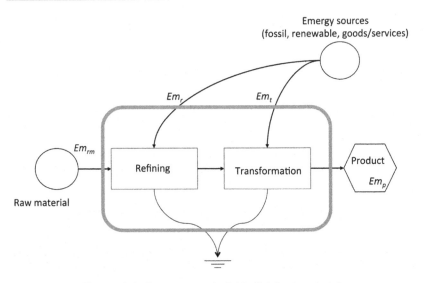

Figure 3.1. *From raw material to finished material*

3.1.1. *Direct production: from raw material to product*

Figure 3.1 represents a procedure that does not include recycling. The extraction emergy of the raw material is written as Em_{rm}, the refining process as Em_r and the transformation process (formation of the product) as Em_t. To reinforce these ideas, an aluminum ingot is considered. All emergy can be composed of three parts: renewable, non-renewable and goods/services. If we apply rule 1, the specific emergy of the product Em_p is the sum of the incoming emergies:

$$Em_p = Em_{rm} + Em_r + Em_t \qquad [3.1]$$

For the example considered (aluminum), electricity is vital to its transformation. The transformation emergy Em_t is therefore the product of the quantity of electricity consumed E_{el} by its specific emergy em_{el}.

The unitary emergy value of the product em_p is expressed by:

$$em_p = \frac{Em_{rm} + Em_r + Em_t}{P} \qquad [3.2]$$

with P being the number of products over the duration of the analysis.

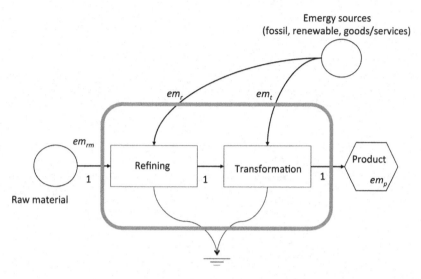

Figure 3.2. *Specific emergy relative to product*

By introducing the unitary energies relative to the product (and not the source; see Figure 3.2), we can rewrite equation [3.2] as:

$$em_p = em_{rm} + em_r + em_t \tag{3.3}$$

where em_{mp}, em_r and em_t are emergies per unit of product. Using these notations, the unitary emergy of transformation em_t is different from the specific emergy of electricity em_{el}.

3.2. Local approach

3.2.1. *Continuous recycling without material loss*

The recycling cycle considered consists of collecting a portion of the product at the end of its life (see Figure 3.3). We are considering a procedure with a recycled mass fraction, written as q.

The lifespan of the product can be much longer than the time required for recycling and transformation. The automobile and real estate industries are examples of applications for aluminum.

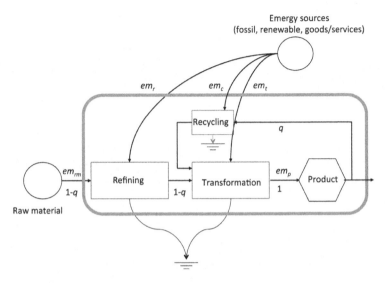

Figure 3.3. *Direct recycling of a product at the end of its life*

Figure 3.3 is said to include implicit time. Emergy is intrinsically linked to an energy/exergy which is dependent on the history of the study (of a temporal graph). This temporal graph must be "unfolded" in order to conduct an emergetic analysis, while this is not required for an energetic analysis. In terms of energy, only the difference in energy consumed between two recycling systems is important, while in terms of emergy, as the point of origin (the source) is the same, the unitary emergy value of a product can only increase from this original value during each cycle. Thus, it is possible to adapt an approach similar to the Lagrangian approach in fluid mechanics, which consists of following the material (the emergy support) during the recycling process.

Figure 3.4 emphasizes the fact that the energy required during a cycle *n* differs from the energy in any preceding cycle. Thus, Rule 4 is not applied; double counting by extending time is not possible in any case.

Let us consider cycle *n*; the specific emergy of product *emₚ(n)* can be written as:

$$em_p(n) = \left(1 - q(n)\right)[em_{rm}(n) + em_r(n)] + \\ q(n)[em_c(n) + em_p(n-1)] + \\ em_t(n)$$

[3.4]

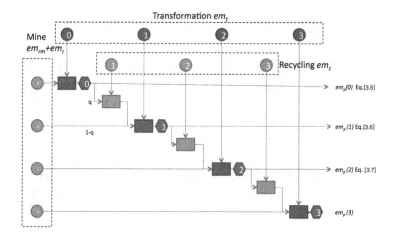

Figure 3.4. *An explicit time description of recycling*

Extraction and refining concern the fraction (1-q), and recycling concerns the portion *q* including the associated specific emergy *em$_c$* required to recycle 100% of the product during a cycle, with transformation being applied to a unit of product.

We obtain a discreet time-recurrent equation (the sampling time of which corresponds to the lifespan of the product). We write *n=0* when the raw material creates the first product and, to simplify the notations, let us suppose that the fraction *q* and the specific emergies per unit of product are constant:

$$em_p(0) = em_{rm} + em_r + em_t \tag{3.5}$$

$$\begin{aligned} em_p(1) &= (1-q)\left[em_{rm} + em_r\right] + q(em_c + em_p(0)) + em_t \\ &= \left[em_{rm} + em_r\right] + qem_c + (1+q)em_t \\ &= em_p(0) + q(em_c + em_t) \end{aligned} \tag{3.6}$$

Whether or not the material has undergone recycling, it has been extracted from the source and refined. We find this emergy content in the term $\left[em_{mp} + em_r\right]$. Emergy due to recycling appears in the term $q\,em_c$. The last term can be broken down into two parts; the first part corresponds to the transformation of the cycle in progress, while the second means that a fraction *q* has already been transformed once. Because emergy is the sum of all resources mobilized directly or indirectly, this term translates the effect of the past for the first recycling.

For the second recycling, we get:

$$
\begin{aligned}
em_p(2) &= (1-q)\left[em_{rm}+em_r\right]+q(em_c+em_p(1))+em_t \\
&= \left[em_{rm}+em_r\right]+q(1+q)em_c+\left(1+q+q^2\right)em_t \\
&= em_p(0)+q(1+q)(em_c+em_t)
\end{aligned}
\tag{3.7}
$$

The cumulative effect of the recycling appears at two levels in the specific emergy of the product: a portion q and a portion q^2 have been recycled and transformed once and twice, respectively.

Using these assumptions, we get the specific emergy of the product in cycle n (for $n \geq 1$):

$$
\begin{aligned}
em_p(n) &= (1-q)\left[em_{rm}+em_r\right]+q(em_c+em_p(n-1))+em_t \\
&= em_p(0)+q(1+q+...+q^{n-1})[em_c+em_t]
\end{aligned}
\tag{3.8}
$$

The sum appears of a geometric progression q, from the first term 1, to the last term q^{n-1}.

$$
em_p(n) = em_p(0) + q\frac{1-q^n}{1-q}[em_c + em_t]
\tag{3.9}
$$

Equation [3.9] reflects the fact that the more a product is recycled, the more its unit emergy value increases. The unit emergy value of a product represents the history of the exergy "invested" in it by nature. It is "the cost we pay to Nature to produce work" [CAR 24]; we might say that the increase in unit emergy value of a product during successive recycling cycles is a measurement of the "cost" paid by nature. This is a counterintuitive fact: the lower the unit emergy value of a product, the more it "weighs" on its environment. Consequently, the higher the unit emergy value of a product, the more "harmful" it will be. Equation [3.9] clearly shows that the minimal value is that of the source and of the refining process $em_p(0)$: thus we should use raw material instead of recycled material. However, the hypotheses used to obtain this equation assume that em_{mp} and em_r are constant. We might think that as the sources, i.e. the mines, are exhausted, extraction and refining will become more difficult, and consequently that the associated energetic (emergetic) resources will become greater. Thus, if the values of em_{mp} and

em_r are large in cycle n, it is entirely possible that for a triplet ($em_{rm}(i)$, $em_r(i)$, $em_c(i)$) with $i<n$, the unit emergy value of the recycled product will become less than the sum of the emergies of the raw material and its refining in cycle n considered $em_{rm}(n)+em_r(n)$. This aspect, then, restores all the relevance of emergy assessment. We would also add that technological breakthroughs and environmental restrictions on emissions may also lead to the reevaluation of the unit emergy values $em_{rm}(n)$, $em_r(n)$, $em_t(n)$, or $em_c(n)$.

The unit emergy value [3.9] can be rewritten as follows:

$$em_p(n) = em_p(0) + q\Psi[em_c + em_t]$$

By introducing the corrective factor Ψ, such that:

$$\Psi = \frac{1-q^n}{1-q} \qquad\qquad [3.10]$$

For a recycled fraction of a material of 100%, it is no longer possible to define an emergy content without knowing the number of cycles carried out. Conversely, for a recycled fraction of up to 80%, the correction factor tends toward its asymptote for a number of cycles lower than 10 (see Figure 3.5). If the asymptotic value is attained, then only one additional piece of information is necessary.

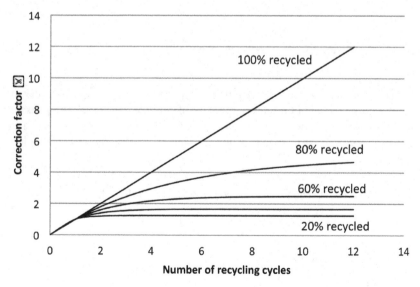

Figure 3.5. *Corrective factor to be introduced during recycling cycles*

3.2.2. *Continuous recycling with material loss during recycling*

Before we consider the general case, let us consider a loss during recycling, such as a rejection during sorting. We will write this loss as p_r and assume it to be proportional to the incoming mass; that is, fraction q (see Figure 3.6).

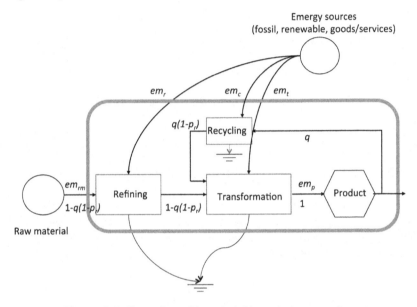

Figure 3.6. *Recycling with material loss during recycling*

The outgoing mass of the recycling is expressed by $q(1 - p_r)$. In order to be able to reason about a unitary quantity of product, we must "compensate for" this loss sustained by the raw material coming from the source, i.e. $1 - q(1 - p_r)$ so as to obtain a unit of product exiting the transformation stage.

$$em_p(n) = \left(1 - q(n)(1 - p_r(n))\right)[em_{rm}(n) + em_r(n)] +$$
$$q(n)\left[em_c(n) + em_p(n-1)\right] + em_t(n) \qquad [3.11]$$

The emergy entering the recycling process is

$$q(n)[em_c(n) + em_p(n - 1)].$$

By considering unit emergy values, the fraction recycled, and the loss as independent of the cycle, we have:

$$em_p(n) = (1 - q(1 - p_r))[em_{rm} + em_r] +$$
$$q[em_c + em_p(n-1)] + em_t \qquad [3.12]$$

With a first term of:

$$em_p(0) = em_{rm} + em_r + em_t \qquad [3.13]$$

To obtain a unit of product, we must use raw material. As in equation [3.9], we obtain the unit emergy value of the product in cycle n (for $n \geq 1$):

$$em_p(n) = em_p(0) + q\frac{1-q^n}{1-q}[em_c + em_t + p_r(em_{rm} + em_r)] \qquad [3.14]$$

The reasoning of geometric progression $q(1 - p_r)$ shows the recycled fraction as well as the loss p_r. The unitary emergy of the recycled product contains the history of losses.

3.2.3. Continuous recycling with material loss during transformation

Figure 3.7 represents a recycling procedure with material loss during transformation.

Using p_t to represent the loss during transformation, the mass balance for a unit of product without loss during recycling requires the quantity of raw material to be equal to $\frac{1}{1-p_t} - q$. As previously stated, we write the specific unitary energy in cycle n as:

$$em_p(n) = \left(\frac{1}{1-p_t(n)} - q(n)\right)[em_{rm}(n) + em_r(n)] +$$
$$q(n)[em_c(n) + em_p(n-1)] + \left(\frac{1}{1-p_t(n)}\right)em_t(n) \qquad [3.15]$$

With a first term of:

$$em_p(0) = \frac{1}{1-p_t}[em_{rm} + em_r + em_t] \qquad [3.16]$$

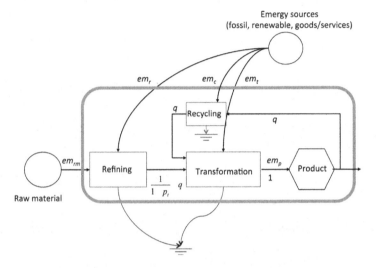

Figure 3.7. *Recycling with a material loss during transformation*

The unit emergy value of the product in cycle n can be written as:

$$em_p(n) = em_p(0) + \frac{q^{n+1} - q}{q - 1}(em_p(0) + em_c - em_{rm} - em_r) \qquad [3.17]$$

3.2.3.1. *Example: aluminum*

Buranakarn [BUR 98, p. 60] provides the values applicable to the recycling of aluminum (see Tables 3.1 and 3.2).

Element		Unit/year	Quantity	Unitary solar emergy	Emergy
				seJ/unit	seJ/year
1	Aluminum (ore)	g	4.17E+11	1.17E+10	4.88E+21
2	Electricity	J	1.08E+15	1.74E+05	1.88E+20
3	Work	$	2.09E+07	1.15E+12	2.40E+19
Annual quantity		g	4.00E+11	1.27E+10	5.08E+21

Table 3.1. *Emergy table for annual aluminum production in the United States*

Element	Unit/year	Quantity	Unitary solar emergy seJ/unit	Emergy seJ/year
1 Aluminum (ore)	g	1.25E+11	1.17E+10	4.88E+21
2 Electricity	J	1.08E+15	1.74E+05	1.88E+20
3 Work	$	2.09E+07	1.15E+12	2.40E+19
4 Recycled aluminum	g	2.29E+11	1.17E+10	2.68E+21
5 Waste aluminum	g	6.25E+10	1.17E+10	7.31E+20
6 Collection	g	2.29E+11	2.51E+08	5.75E+19
7 Separation	g	2.29E+11	8.24E+06	1.89E+18
8 Transport	Ton/mile	2.82E+07	9.65E+11	2.72E+19
Annual quantity	**g**	**4.00E+11**	**To be calculated**	**To be calculated**

Table 3.2. *Emergy table for annual aluminum production in the United States [BUR 98, p. 60]*

With the notations introduced, we have: $em_{rm} + em_r = 1.22E+10$ seJ/g alu; $em_t = 5.30E+08$ seJ/g alu.

This allows us to obtain the unit emergy value of aluminum:

$em_c = 2.97E+08$ seJ/g alu[1], $p_t = 4\%$, $q = 72.87\%$.

Figure 3.8 shows the evolution of the unit emergy value of aluminum.

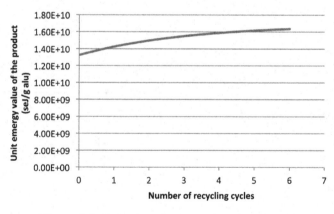

Figure 3.8. *Evolution of unit emergy of aluminum*

---

1 For this calculation, we must take the emergy of lines 6 + 7 + 8 and weight this sum, and since by definition em_c must correspond to 100% and the fraction q is 72.87%, we must introduce an additional correction.

3.2.4. *Discontinuous recycling*

To approximate the reality of recycling, we must consider that materials which have been recycled *i* times are mixed with materials recycled *j* times.

First, we must study the evolution of unit emergy values in recycling without renewal (unlike the cases presented above). Raw material initiates cycles and material is then released (see Figure 3.9). It is first necessary to write the mass balance between two recycling cycles, and then the total energy, in order to deduce the unit emergy value. The system of equations is given with the initial conditions; see equation [3.18].

$$
\begin{cases}
\begin{cases}
m_p^{rc}(n) = q(1 - p_c)(1 - p_t)\, m_p^{rc}(n - 1) \\
Em_p^{rc}(n) = q\, m_p^{rc}(n - 1)[em_p^{rc}(n - 1) + em_c + (1 - p_c)em_t] \\
em_p^{rc}(n) = \dfrac{Em_p^{rc}(n)}{m_p^{rc}(n)}
\end{cases} \\
\begin{cases}
m_p^{rc}(0) = 1 \\
Em_p^{rc}(0) = \dfrac{1}{1 - p_t}(em_{mp} + em_r + em_t) \\
em_p^{rc}(0) = \dfrac{Em_p^{rc}(0)}{m_p^{rc}(0)}
\end{cases}
\end{cases}
\qquad [3.18]
$$

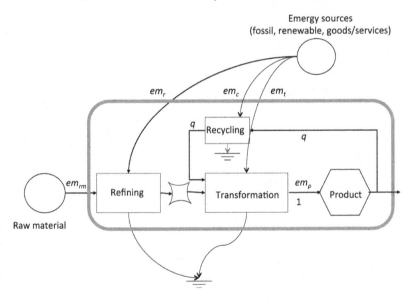

Figure 3.9. *Recycling in release cycle*

The unit emergy value $em_p(n)$ thus obtained is not dependent on mass and corresponds to the emergy of a material that has undergone n cycles.

$$em_p(n) = \frac{em_p(n-1) + em_c + (1-p_c)em_t}{(1-p_c)(1-p_t)}$$ [3.19]

Let us consider again the example of aluminum; the first elements of the progression of equation [3.19] are given:

$em_p^{rc}(0) = 1.33E+10$ seJ/g alu, $em_p^{rc}(1) = 1.47E+10$ seJ/g alu,
$em_p^{rc}(2) = 1.61E+10$ seJ/g alu, $em_p^{rc}(3) = 1.77E+10$ seJ/g alu.

The first elements of the progression of equation [3.17] are given:

$em_p(0) = 1.33E+10$ seJ/g alu, $em_p(1) = 1.42E+10$ seJ/g alu,
$em_p(2) = 1.50E+10$ seJ/g alu, $em_p(3) = 1.55E+10$ seJ/g alu.

In both cases, the initial value is indeed the same; however, all the other values are different. In the recycling diagram shown in Figure 3.7, to which equation [3.17] corresponds, the recirculating mass breaks down as follows:

– for the first transformation, i.e. without recycling, we introduce $\left(\frac{1}{1-p_t}\right)$ as raw material and obtain one unit of product, i.e. $\left(\frac{1}{1-p_t}\right)(1-p_t)$. In other words, 100% of the product comes from the raw material;

– for the first recycling, fraction q is thus composed of 100% raw material. This fraction now undergoes its first recycling. The mass entering the transformation stage is composed of $\left(\frac{1}{1-p_t}\right) - q$ of raw material and q of material recycled once. The mass exiting the transformation stage is composed of $\left(\frac{1}{1-p_t} - q\right)(1-p_t)$ of raw material and $q(1-p_t)$ of material recycled once;

$$em_p(1) = \left(\frac{1}{1-p_t} - q\right)(1-p_t)em_p^{rc}(0) + q(1-p_t)em_p^{rc}(1)$$

– for the second recycling, fraction q is broken down into two terms: $q\left(\frac{1}{1-p_t} - q\right)(1-p_t)$ undergoes its first recycling, and $q(q(1-p_t))$ undergoes its second recycling. The outgoing mass is composed of three terms: raw material $\left(\frac{1}{1-p_t} - q\right)(1-p_t)$, the part of the mass that has undergone one recycling $q\left(\frac{1}{1-p_t}-q\right)(1-p_t)(1-p_t)$, and the part of the mass that has undergone two recycling cycles $q\left(q(1-p_t)\right)(1-p_t)$;

$$em_p\,(2) = \left(\frac{1}{1-p_t}-q\right)(1-p_t)em_p^{rc}(0) +$$
$$q\left(\frac{1}{1-p_t}-q\right)(1-p_t)(1-p_t)em_p^{rc}(1) +$$
$$q\left(q(1-p_t)\right)(1-p_t)em_p^{rc}(2)$$

– for the third recycling, the fraction q is broken down into three terms:

- first recycling: $\quad q\left(\frac{1}{1-p_t}-q\right)(1-p_t)$

- second recycling: $\quad q^2\left(\frac{1}{1-p_t}-q\right)(1-p_t)^2$

- third recycling: $\quad q^3\,(1-p_t)^2$

Using $x(i)$ to write the mass fraction exiting transformation and having undergone i recycling cycles, the outgoing mass is thus composed of four terms:

- raw material: $\quad x(0) = \left(\frac{1}{1-p_t}-q\right)(1-p_t)$

- first recycling: $\quad x(1) = q\left(\frac{1}{1-p_t}-q\right)(1-p_t)^2$

- second recycling: $\quad x(2) = q^2\left(\frac{1}{1-p_t}-q\right)(1-p_t)^3$

- third recycling: $\quad x(3) = q^3\,(1-p_t)^3$

The unit emergy value $em_p\,(3)$ is written with the mass fractions $x(i)$:

$$em_p\,(3) = \textstyle\sum_{i=0}^{3} x(i)em_p^{rc}(i)$$

It should be noted that because the sources are assumed to be independent, and the emergy of a product being the sum of its sources, the emergy function is *additive* in comparison to the sources.

The most general case of recycling means considering a stored quantity of recycled materials with a fraction $x(i)$ having undergone i recycling cycles (see Figure 3.10).

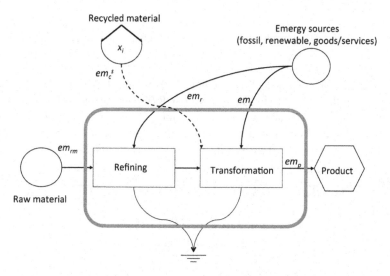

Figure 3.10. *Recycled material coming from storage*

The unit emergy value of the product is thus obtained from:

$$em_p(n) = \left(\frac{1}{1-p_t} - q\right)(em_{rm} + em_r) + q\,em_c^o(n) + \frac{1}{1-p_t}\,em_t \quad [3.20]$$

The unit emergy value coming from the stored recycled material, written as em_{cs}, is considered at the same time as the unit emergy value of the product is assessed. Conversely, the calculation of $em_c^o(n)$ is itself a consequence of the history/pathway taken by the material.

$$em_c^o(n) = \sum_{i=0}^{I} x(i)\,em_p^{rc}(i) \quad [3.21]$$

With I the maximum number of cycles contained by a fraction of material exiting storage, with $I \le n$.

3.2.4.1. *Example: aluminum*

Bertram *et al.* [BER 09, p. 652] provided global aluminum production rates (see Table 3.3).

Year	1950	1960	1970	1980	1990	2000	2010
Percentage recycled	18%	20%	22%	25%	28%	35%	31%

Table 3.3. *Recycling rate for aluminum [BER 09, p. 652]*

NOTE.– The percentage of recycling drops between 2000 and 2010, but this does not mean a drop in the quantity recycled; rather, aluminum production has continued to increase, and it appears difficult for recycling capacity to keep up.

The mass fraction q is thus taken at its current value (31%, quite a different value from the one given by Buranakarn [BUR 98], who evaluated only for the United States). To be able to calculate the unit emergy value of a product (an ingot exiting the transformation unit), it is necessary to have an estimate of the distribution of the mass fraction x (i). An example is given in Table 3.4.

Recycling	6 times	5 times	4 times	3 times	2 times	1 time
Mass fraction $x(i)$	1%	2%	7%	15%	25%	50%
Unitary emergy seJ/g alu	$em_p^{SR}(6)$ 2.27E+10	$em_p^{SR}(5)$ 2.09E+10	$em_p^{SR}(4)$ 1.93E+10	$em_p^{SR}(3)$ 1.77E+10	$em_p^{SR}(2)$ 1.61E+10	$em_p^{SR}(1)$ 1.47E+10

Table 3.4. *Distribution of mass fraction*

By applying equation [3.21], we obtain the value of the unitary emergy of the product equal to 1.52E+10 seJ/alu upon exiting the transformation stage.

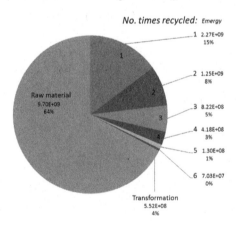

Figure 3.11. *Emergy of an aluminum ingot exiting the transformation phase. For a color version of the figure, see www.iste.co.uk/lecorre/emergy.zip*

The emergy contribution of the raw material and of its refining corresponds to 64% of the unit emergy value of the ingot (and 69% by mass) (see Figure 3.11). The current transformation represents only 4% of the unit emergy value of the product. For recycled material we have the following pairs (mass percentage; emergy percentage):

– 1 time (15.5%; 14.9%),

– 2 times (7.7%; 8.2%),

– 3 times (4.6%; 5.4%)

– 4 times (2.1%; 2.7%),

– 5 times (0.6%; 0.8%) et

– 6 times (0.4%; 0.5%)

We can say that the contribution in terms of emergy of each type of recycled material corresponds to its mass proportion. It is important to note that in the case studied,

$$em_t = 4.3\% \; (em_{rm} + em_r) \text{ and } em_c = 2.4\% \; (em_{rm} + em_r)$$

(and still less if we have losses p_t).

3.2.5. Comparison with a global approach

The global approach differs from the previous approach in that temporal aggregation occurs in the sources. Again using Figure 3.4, we can perform the sum of the emergy sources (over a given period of time; for example the period corresponding to n recycling cycles) and develop rules for emergy calculation. Four sources are considered:

1) a supply of raw material;

2) refining;

3) various transformations;

4) various recycling cycles.

3.2.5.1. Continuous recycling without material loss

We will consider ourselves to be in the same hypothetical context as in section 3.2.1 (see Figure 3.3). The total emergy of the raw material is the sum of the emergies of each supply. The first supply corresponds to the unitary emergy em_{mp} and the n following supplies to the fraction $\left((1 - q)\, em_{rm}\right)$. We obtain the total emergy of the raw materials $Em_{rm}(n)$ by:

$$Em_{rm}(n) = \sum_{i=0}^{n} Em_{rm}(i) = \left(1 + n\,(1 - q)\right)em_{rm} \qquad [3.22]$$

In the same way, we obtain the total emergy of the refining process $Em_r(n)$ by:

$$Em_r(n) = \sum_{i=0}^{n} Em_r(i) = \left(1 + n\,(1-q)\right)em_r \qquad [3.23]$$

The total transformation emergy $Em_t(n)$ is expressed as:

$$Em_t(n) = \sum_{i=0}^{n} Em_t(i) = (1+n\,)em_t \qquad [3.24]$$

And finally, the total recycling emergy $Em_c(n)$ is given by:

$$Em_c(n) = \sum_{i=1}^{n} Em_c(i) = (q\,n\,)em_c, \qquad [3.25]$$

being careful to note the starting index, which for recycling is $i=1$, while for the other sources it is $i=0$.

Figure 3.12 shows the classic (implicit-time) emergetic diagram. The source values are calculated using equations [3.22]–[3.25].

The application of the rules of emergy (see sections 1.1 and 1.3) gives the emergy of n products Em_p:

$$Em_p(n) = Em_{rm} + Em_r + Em_t + Em_c \qquad [3.26]$$

The average unitary emergy of n products $\langle em_p(n)\rangle$ is:

$$\langle em_p(n)\rangle = \frac{(1+n\,(1-q))(em_{rm}+em_r)+(1+n\,)em_t+(q\,n\,)em_c}{n}$$

When the number of recycling cycles increase, we get the asymptotic behavior given by:

$$\lim_{n\to\infty}\langle em_p(n)\rangle = (1-q)(em_{rm} + em_r) + em_t + q\,em_c \qquad [3.27]$$

The global approach equation [3.27] can be compared to the Lagrangian approach equation [3.9]:

$$\lim_{n\to\infty} em_p(n) = em_p(0) + \frac{q}{1-q}(em_c + em_t) \qquad [3.28]$$

Thus we have the following result:

$$\lim_{n\to\infty}\langle em_p(n)\rangle \neq \lim_{n\to\infty} em_p(n)$$

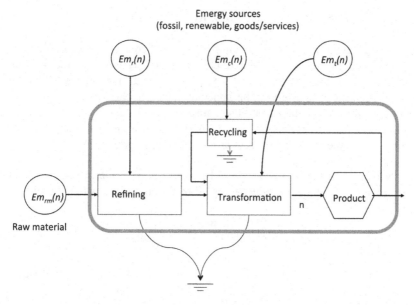

Figure 3.12. *Recycling without loss of mass using a global approach*

3.2.5.2. *Conclusion*

The values of the asymptotic behavior of the unit emergy value and its average are different. The global approach suggests that n products are available at the end of the considered time period, but this is false since only a single product is available after each cycle.

Equation [3.27] is clearly incorrect. In fact, the minimum value of the emergy of a product, whatever it may be, is obviously equal to or greater than the sum of the input emergies during its first development: $em_{mp} + em_r + em_t$, in the context of the hypotheses given here. The application of rule 4 thus makes temporal aggregation impossible, generally speaking. It should be noted that rule 4 also makes spatial aggregation impossible (i.e. the assembling of blocks).

Conversely, we can see that if the recycled fraction is equal to zero (no recycling), the two limits are equal. The global approach and the Lagrangian approach correspond to one another.

3.3. Recycling with losses of quality and material

In the previous section, recycling was considered as a single unit. For metals, multiple stages are required: collection and separation, shredding and sorting, and fusing and formulation of the compound that emerges from the recycling project. Waste may be generated at each stage (corresponding to material loss); each stage may also cause a loss of quality of the outgoing component. Additives may be used to compensate for material and/or quality losses due to the recycling process as a whole (see Figure 3.13).

Since transformity is the ratio between source emergy and product exergy, a degradation of product exergy directly influences its transformity.

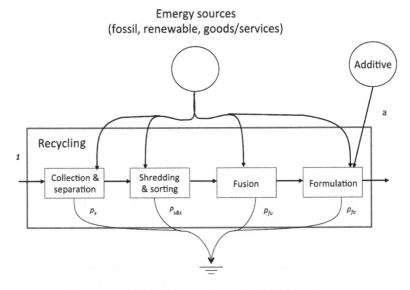

Figure 3.13. *Breakdown of recycling into four stages*

The mass balance of the recycling part includes various losses and a possible contribution of an additive. The outgoing mass m_c^o of the recycling process is:

$$m_c^o = \left((1 - p_s)(1 - p_{s\&s})(1 - p_{fu}) + a\right)(1 - p_{fo})$$

with $p_s, p_{s\&s}, p_{fu}, p_{fo}$ respectively representing the losses during separation, shredding (and shorting), fusion, and formulation, and a the quantity of additive.

The shredding stage consists of destroying the bonds in the material being recycled. It is irreversible, and is thus associated with a loss of exergy. Likewise, the stages of fusion and formulation are intrinsically irreversible.

However, the unit emergy value, or transformity, of a product is the ratio between the emergy of the sources required for the constitution of this product and its exergy. Material losses cause the source emergies to increase (see section 3.2), resulting in increased transformity. Degradations of material quality also cause a reduction of the denominator and are manifested in increased transformity. Both mass loss and irreversibility act on the numerator and the denominator.

3.3.1. Quantification of exergy losses

Szargut et al. [SZA 05] proposed a theoretical context for the evaluation of the chemical exergy of materials, ex_m. This is composed in part of a free enthalpy of formation Δg_f, and the sum of the chemical exergies of the elements (elementary types of molar fraction y_i) composing the material, as well as of a term translating the irreversible natures of the elements mixture:

$$ex_m = \Delta g_f + \Sigma_i y_i \, ex_{ch,i} + RT^0 \Sigma_i y_i \ln(\alpha_i) \qquad [3.29]$$

Activity α_i is equal to 1 if there is no interaction between elements. The stronger the interactions, the smaller the activity[2] (see Castro et al. [CAS 04]). For an ideal mixture, the activity can be replaced by the molar fraction y_i [FIN 97].

3.3.2. Exergy losses related to recycling

Recycling takes place in four stages (see Figure 3.14):

– collection and separation;

– shredding and sorting;

– fusion (for materials requiring this stage);

– formulation.

---

2 The presence of contaminants in the recycled product increases disorder caused by interactions between elements.

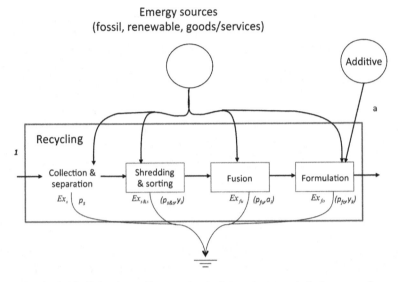

Figure 3.14. *Overview of losses (material and emergy) during recycling*

3.3.2.1. *Collection and separation*

The loss of exergy by a material during recycling is proportional to its losses of mass p_s:

$$\Delta Ex_s = p_s\, ex_s \qquad\qquad [3.30]$$

3.3.2.2. *Shredding and sorting*

The process of shredding destroys atomic bonds, and therefore this stage introduces irreversible changes that are manifested as a destruction of exergy. Ignatenko *et al.* [IGN 07] proposed expressing this destruction by introducing the molar fraction of the constituents into equation [3.29] on the basis that there is no change of nature. The exergy loss ΔEx_d at this stage is the difference between the input Ex_d^i and the output Ex_d^o.

$$\Delta Ex_{s\&s} = Ex_{s\&s}^i - Ex_{s\&s}^o \qquad\qquad [3.31]$$

The incoming exergy corresponds to the output of the collection stage weighted by the material loss during this stage:

$$Ex_{s\&s}^i = (1 - p_{s\&s})\left[\sum_i y_i\, ex_{ch,i}\right]$$

The outgoing exergy is obtained by applying equation [3.28], keeping the losses:

$$Ex^o_{s\&s} = (1 - p_s)(1 - p_{s\&s})\left[\sum_i y_i \, ex_{ch,i} + RT^0 \sum_i y_i \ln(y_i)\right]$$

Finally, we obtain the exergetic loss:

$$\Delta Ex_{s\&s} = (1 - p_s)\, p_{s\&s} \left[\sum_i y_i \, ex_{ch,i}\right] - (1 - p_s)(1 - p_{s\&s})[RT^0 \sum_i y_i \ln(y_i)]$$

3.3.2.3. Fusion

During the fusion stage, irreversibilities depend on the activity of bonds, through the term $RT^0 \sum_i y_i \, ln(\alpha_i)$. By denoting those irreversibilities as ΔEx_{fu}, we get:

$$\Delta Ex_{fu} = Ex^i_{fu} - Ex^o_{fu} \tag{3.32}$$

Similarly to the preceding stage, the incoming exergy Ex^i_{fu} is the output of shredding and sorting:

$$Ex^i_{fu} = (1 - p_s)(1 - p_{s\&s})\left[\sum_i y_i \, ex_{ch,i} + RT^0 \sum_i y_i \ln(y_i)\right].$$

The outgoing exergy takes into account the destruction of exergy and material loss during this stage:

$$Ex^o_{fu} = (1 - p_s)(1 - p_{s\&s})(1 - p_{fu})\left[\sum_i y_i \, ex_{ch,i} + RT^0 \sum_i y_i \ln(\alpha_i)\right]$$

We obtain the exergy loss ΔEx_{fu}:

$$\Delta Ex_{fu} = (1 - p_s)(1 - p_{s\&s})\, p_{fu}\left[\sum_i y_i ex_{ch,i}\right] +$$

$$(1 - p_s)(1 - p_{s\&s})\left[RT^0 \sum_i y_i \ln\ln(y_i)\right]$$

$$-(1 - p_s)(1 - p_{s\&s})(1 - p_{fu})\left[RT^0 \sum_i y_i \ln\ln(\alpha_i)\right]$$

3.3.2.4. *Formulation*

The stage of formulating recycled material may require the introduction of an additive, of fraction a.

$$\Delta Ex_{fo} = Ex_{fo}^i + Ex_a - Ex_{fo}^o \qquad [3.33]$$

As previously, the incoming exergy is the output of the fusion stage:

$$Ex_{fo}^i = (1 - p_s)(1 - p_{s\&s})(1 - p_{fu})[\textstyle\sum_i y_i \, ex_{ch,i} + RT^0 \sum_i y_i \ln(\alpha_i)]$$

The exergy of the additive is weighted by its molar fraction:

$$Ex_a = a \, \textstyle\sum_k y_k ex_{ch,k}$$

Finally, the outgoing exergy takes into account contributions by the additive, destructions of exergy, and losses of mass:

$$Ex_{fo}^o = \left((1 - p_s)(1 - p_{s\&s})(1 - p_{fu}) + a\right)\left(1 - p_{fo}\right)[\textstyle\sum_j y_j \, ex_{ch,j} + RT^0 \sum_j y_j \ln(\alpha_j)]$$

Table 3.5 summarizes the different losses according to the stages of recycling:

Stage		
Collection and separation	Material loss	ΔEx_s
Shredding and sorting	Material loss and irreversibility	$\Delta Ex_{s\&s}$
Fusion	Material loss and irreversibility	ΔEx_{fu}
Formulation	Material loss and irreversibility	ΔEx_{fo}

Table 3.5. *Summary of exergy losses according to stages of the recycling process*

3.3.3. *Recycling with additive (with no losses during refining and transformation)*

Figure 3.15 shows a recycling process that includes an additive. This case is similar to the one presented above (referred to as being "without renewal").

We introduce the following two notations:

$$p_{pr} = 1 - (1 - p_s)(1 - p_{s\&s})(1 - p_{fu}) \qquad [3.34]$$

$$p_{c,a} = 1 - \frac{(1-p_p+a)(1-p_{fo})}{(1+a)} \qquad [3.35]$$

with p_{pr} and $p_{c,a}$ representing partial losses occurring during formulation and after formulation taking into account the use of an additive, respectively.

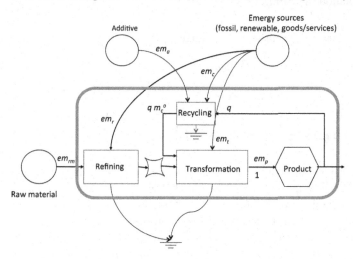

Figure 3.15. *Recycling with additive*

Table 3.6 specifies the incoming masses in unitary operations per unit of product.

	Mass (g)	Unitary emergy (seJ/g)	Total emergy contributed by unitary operation (seJ)
Raw material	1	em_{mp}	em_{mp}
Refining	1	em_r	em_r
Collection and separation	q	em_s	$q\,em_s$
Shredding and sorting	$q(1-p_s)$	$em_{s\&s}$	$q(1-p_s)\,em_{s\&s}$
Fusion	$q(1-p_s)(1-p_{s\&s})$	em_{fu}	$q(1-p_s)(1-p_{s\&s})\,em_{fu}$
Formulation	$q(1-p_p+a)$	em_{fo}	$q(1-p_p+a)\,em_{fo}$
Recycling output	$q(1-p_{c,a})(1+a)$	$em_{c,a}$	$q(1-p_{c,a})(1+a)\,em_{c,a}$

Table 3.6. *Description of various emergies according to stage*

The emergetic balance associated with the formulation stage is the sum of the additive $q\,a\,em_a$ and the emergy contribution related to formulation $q\left(1 - p_{pr} + a\right)em_{fo}^T$; see equation [3.36]. Taking into account the outgoing mass $q\left(1 - p_{pr} + a\right)$, we define the specific emergy contributed during the formulation stage em_{fo}:

$$Em_{fo} = q\left(aem_a + \left(1 - p_{pr} + a\right)em_{fo}^T\right)$$
$$= q\left(1 - p_{pr} + a\right)em_{fo}$$

[3.36]

with

$$em_{fo} = \frac{a}{\left(1 - p_{pr} + a\right)}\,em_a + em_{fo}^T$$

[3.37]

The material being recycled may require various additives, so the part a may vary with the number of recycling cycles. For the first recycling, the total output emergy of recycling, written as $Em_c^o(1)$, is the sum of the total emergy carried by the material entering the recycling process $q[em_{rm} + em_r + em_t]$, the total emergy associated with collection and separation qem_s, the total emergy associated with shredding and sorting $q\left(1 - p_s\right)em_{s\&s}$, the total emergy associated with fusion $q\left(1 - p_s\right)\left(1 - p_{s\&s}\right)em_{fu}$, and the emergy necessary for the formulation stage:

$$Em_c^o(1) = q\left(1 - p_{c,a}(1)\right)\left(1 + a(1)\right)em_{c,a}(1)$$
$$= q[em_{rm} + em_r + em_t + em_s + \left(1 - p_s\right)em_{s\&s} +$$
$$\left(1 - p_s\right)\left(1 - p_{s\&s}\right)em_{fu} + \left(1 - p_{pr} + a(1)\right)em_{fo}]$$

[3.38]

Using the previous notation, the specific emergy coming out of the recycling process, $em_{c,a}^o$, is the sum of the incoming emergies:

$$em_{c,a}^o(1) = \frac{1}{\left(1 - p_{c,a}(1)\right)\left(1 + a(1)\right)}\left[em_{rm} + em_r + em_t\right] + em_c^T(1)$$ [3.39]

where the specific emergy for the recycling process, em_c^T, is defined for the first recycling by:

$$em_c^T(1) = \frac{em_s(1 - p_s) + em_{s\&s} + (1 - p_s)(1 - p_{s\&s})\,em_{fu} + (1 - p_{pr} + a(1))\,em_{fo}}{(1 - p_{c,a}(1))(1 + a(1))}$$

[3.40]

The transformity $Tr_{c,a}^o$ is defined in relation to the exergy $ex_{c,a}^o$ (while the unit emergy value $em_{c,a}^o$ refers to the mass):

$$Tr_{c,a}^o(1) = \frac{em_{c,a}^o(1)}{ex_{c,a}^o(1)}$$ [3.41]

As in the first recycling, we obtain the recurrent equation for the specific emergy output of recycling:

$$em_{c,a}^o(n) = \frac{1}{(1-p_{c,a}(n))(1+a(n))}\left[em_{c,a}^o(n-1) + em_t\right] + em_c^T(n)$$ [3.42]

With the emergy contributed during the recycling process $em_c^T(n)$ defined by:

$$em_c^T(n) = \frac{em_s(1-p_s) + em_{s\&s}+(1-p_s)(1-p_{s\&s})\,em_{fu}+(1-p_{pr}+a(n))\,em_{fo}}{(1-p_{c,a}(n))(1+a(n))}$$ [3.43]

the transformity of the output material from recycling is:

$$Tr_{c,a}^o(n) = \frac{em_{c,a}^o(n)}{ex_{c,a}^o(n)}$$ [3.44]

The unitary emergy for the second cycle is written as:

$$em_{c,a}^o(2) = \frac{1}{(1-p_{c,a}(2))(1+a(2))}\left[em_{c,a}^o(1) + em_t\right] + em_c^T(2)$$ [3.45]

Developing equation [3.45], we get:

$$em_{c,a}^o(2) = \frac{1}{(1-p_{c,a}(2))(1+a(2))}\left[\frac{1}{(1-p_{c,a}(1))(1+a(1))}\left[em_{rm} + em_r + em_t\right] + em_{c,e}(1) + em_t\right] + em_c^T(2)$$

$$em_{c,a}^o(2) = \frac{1}{\left(1-p_{c,a}(2)\right)\left(1+a(2)\right)}\frac{1}{\left(1-p_{c,a}(1)\right)\left(1+a(1)\right)}\left[em_{rm} + em_r\right]$$

$$+ \frac{1}{\left(1-p_{c,a}(2)\right)\left(1+a(2)\right)}\frac{1}{\left(1-p_{c,a}(1)\right)\left(1+a(1)\right)}\left[em_t\right]$$

$$+ \frac{1}{\left(1-p_{c,a}(2)\right)\left(1+a(2)\right)}em_t + \frac{1}{\left(1-p_{c,a}(2)\right)\left(1+a(2)\right)}em_c^T(1) + em_c^T(2)$$

With:

$$em_c^T(2) = \frac{em_s(1-p_s) + em_{s\&s}+(1-p_s)(1-p_{s\&s})\,em_{fu}+(1-p_{pr}+a(2))\,em_{fo}}{(1-p_{c,a}(2))(1+a(2))}$$ [3.46]

Assuming that all the values are time-constant, we obtain the specific emergy $em_{c,a}^o(n)$ of the cascade of n recycling cycles:

$$
\begin{aligned}
em_{c,a}^o(n) \quad &= \prod_{i=1}^n \frac{1}{(1-p_{c,a})(1+a)} [em_{rm} + em_r] \\
&+ \sum_{i=1}^n \prod_{j=i}^n \frac{1}{(1-p_{c,a})(1+a)} em_t \\
&+ \sum_{i=1}^{n-1} \prod_{j=i+1}^n \frac{1}{(1-p_{c,a})(1+a)} em_c^T(i) + em_c^T(n)
\end{aligned}
$$

It is possible to rewrite $em_{c,a}^o(n)$ as:

$$
\begin{aligned}
em_{c,a}^o(n) \quad &= \left[\frac{1}{(1-p_{c,a})(1+a)} \right]^n [em_{rm} + em_r] \\
&+ \sum_{i=1}^n \left[\frac{1}{(1-p_{c,a})(1+a)} \right]^i em_t \\
&+ \sum_{i=1}^n \left[\frac{1}{(1-p_{c,a})(1+a)} \right]^{n-i} em_c^T(i)
\end{aligned}
$$

The specific emergy of the product $em_p(n)$ after its transformation (being put into its final form) is:

$$
em_p(n) = em_t(n) + em_{c,a}^o(n)
$$

Using equation [3.42], we can define the specific emergy contribution $em_p^T(n)$ received by the recycled product (rp) during one cycle by:

$$
em_{rp}^T(n) = \frac{1}{(1-p_{c,a}(n))(1+a(n))} [em_{rm} + em_r] + em_c^T(n) \qquad [3.47]
$$

Jamali-Zghal [JAM 14] defines the average transformity of recycled products by:

$$
\langle Tr_{rp}\rangle(n) = \frac{(em_{rm}+em_r)+\sum_{i=1}^n \frac{1}{(1-p_{c,a}(i))(1+a(i))} em_t+\sum_{i=1}^n em_c^T(i)}{\overline{ex}_{mr}+\sum_{i=1}^n \langle ex_p\rangle(i)} \qquad [3.48]
$$

with $\langle ex_p\rangle(i)$ the specific exergy of a product containing material recycled i times.

This definition of average transformity corresponds to the global approach described in section 3.2.5. It is possible to factorize equation [3.48] as:

$$\langle Tr_{rp}\rangle(n) = \frac{(em_{rm}+em_r)\left[1+\sum_{i=1}^{n}\frac{1}{(1-p_{c,a}(i))(1+a(i))}\left(\frac{em_t}{em_{rm}+em_r}\right)+\sum_{i=1}^{n}\left(\frac{em_c^T(i)}{em_{rm}+em_r}\right)\right]}{\overline{ex}_{rm}\left[1+\sum_{i=1}^{n}\frac{\langle ex_p\rangle(i)}{\langle ex_{rm}\rangle}\right]} \quad [3.49]$$

$$\langle Tr_{rp}\rangle(n) = \langle Tr_{rm}\rangle\frac{\left[1+\sum_{i=1}^{n}\frac{1}{(1-p_{c,a}(i))(1+a(i))}\left(\frac{em_t}{em_{rm}+em_r}\right)+\sum_{i=1}^{n}\left(\frac{em_c^T(i)}{em_{rm}+em_r}\right)\right]}{\left[1+\sum_{i=1}^{n}\frac{\langle ex_p\rangle(i)}{\langle ex_{rm}\rangle}\right]} \quad [3.50]$$

Three ratios result:

1) the material formation indicator (*MFI*) compares transformation emergy with the emergy of the raw material.

$$MFI(n) = \frac{em_t(n)}{em_{rm}+em_r}$$

The lower this ratio, the more the formation of the product, represented by its specific emergy em_t, is weak compared to the emergy of the refined raw material, and the more the product can be recycled;

2) the recyclability indicator (*RI*) is defined as the ratio between the specific emergy required for cycle i and the emergy of the refined raw material;

$$RI(n) = \frac{em_c^T(n)}{em_{mp}+em_r}$$

The lower this ratio, the more it is of interest from an emergy point of view to recycle the material;

3) the quality ratio *(QR)* for a recycled product is defined as the ratio between the exergy of a recycled material and the exergy of a product after refining:

$$QR(n) = \frac{\langle ex_p\rangle(n)}{\langle ex_{mr}\rangle}$$

The quality ratio of a product QR cannot be greater than 1. The closer this ratio is to 1, the more recyclable the material being considered (meaning that few irreversibilities appear during recycling). Jamali-Zghal [JAM 14] has associated this ratio with the idea of eco-design. The more a product mobilizes a material with a high QR value, the more recyclable it is, and thus the less terrestrial resources are mobilized. QR is an indicator that falls within the concept of sustainable development as discussed in the introduction.

We can rewrite the average transformity with the previous ratios:

$$\langle Tr_{rp}\rangle(n) = \langle Tr_{rm}\rangle^{\frac{\left[\sum_{i=1}^{n}\frac{1}{\left(1-p_{c,a}(i)\right)\left(1+a(i)\right)}MFI(i)+\sum_{i=0}^{n}RI(i)\right]}{\sum_{i=0}^{n}QR(i)}}$$ [3.51]

with $RI(0) = 1$.

For a material to be considered suitable within the context of sustainable development; that is, the context of eco-design, it must have the property:

$$\langle Tr_{rp}^{eco}\rangle(n) < \langle Tr_{rm}\rangle$$

This property compares the transformities of materials entering the transformation (formation) process of the product only during the cycle being considered. The recycled material contains a higher quantity of emergy. However, from a "product" perspective, the competition between raw material and recycled material may be perceived on the basis of the average emergy mobilized. Thus, the emergetic value of the recycled material can be broken down into two parts: an "inevitable" part coming from the previous *(n-1)* cycles, and the part that may possibly be substituted as a result of choices made during the current stage *n*.

This implies that:

$$\frac{\left[\sum_{i=1}^{n}\frac{1}{\left(1-p_{c,a}(i)\right)\left(1+a(i)\right)}MFI^{eco}(i)+\sum_{i=0}^{n}RI^{eco}(i)\right]}{\sum_{i=0}^{n}QR(i)} < 1$$

From this we can deduce two inequalities (knowing that ratios can only be positive):

$$\frac{\sum_{i=1}^{n}RI^{eco}(i)}{\sum_{i=0}^{n}QR(i)} < 1$$ [3.52]

$$\frac{\left[\sum_{i=1}^{n}\frac{1}{\left(1-p_{c,a}(i)\right)\left(1+a(i)\right)}MFI^{eco}(i)\right]}{\sum_{i=0}^{n}QR(i)} < 1$$ [3.53]

If the ratios were constant, we would obtain the limit:

$$MFI^{eco} < \frac{\frac{1+n\,QR}{n}}{(1-p_{c,a})(1+a)}$$

To get:

$$MFI^{eco} < (1 - p_{c,a})(1 + a)\left(RQ + \frac{1}{n}\right)$$

For a large n, we get the limit:

$$MFI^{eco}(\infty) = \lim_{n\to\infty} MFI^{eco} = (1 - p_{c,a})(1 + a)QR \qquad [3.54]$$

The interest of the recyclability index RI is clear from a design perspective; when materials are chosen, we can choose the material with the highest recyclability index. Moreover, the limit formation indicator $MFI^{eco}(\infty)$ makes it possible to compare the unitary emergy that must be "invested" in order to recycle a material with the "cost" of a material composed of raw material.

If we know the QR ratio of a material, we can deduce the maximum number of recycling cycles n_{max} by:

$$MFI^{eco} < MFI^{eco}(\infty) + \frac{(1-p_{c,a})(1+a)}{n_{max}}$$

We obtain the maximum number n_{max}:

$$n_{max} = \frac{(1-p_{c,a})(1+a)}{MFI^{eco} - MFI^{eco}(\infty)}$$

3.3.3.1. *Example of application*

Castro *et al.* [CAS 07] have calculated the exergy degradation for a recyclable aluminum alloy with a base of *Al2036* (*Al 96.6%, Cu 2.6%, Si 0.5%*) and containing iron-based impurities. To improve the quality of the recycled aluminum, the recyclable alloy is mixed with iron-free aluminum *Al2036*, which yields a new alloy of lower quality, *Al308* (*Al 96.1%, Cu 2.6%, Fe 0.8%, Si 0.5%*). Using the notations introduced, we have:

– specific exergy of raw material $ex_{mp} = 32.105 \left[\frac{MJ}{kg}\right]$;

– specific exergy of product recycled once $ex_{mr}(1) = 31.893$ $[\frac{MJ}{kg}]$;

– mass fraction of addition (in the case of *Al2036*) $a = 14.05\%$;

– sum of losses during collection and separation, shredding and sorting, and fusion $p_s + p_d + p_{fu} = 10\%$;

– loss during fusion $p_{fu} = 4.6\%$.

The overall loss during the first recycling cycle $p_{c,a}(1)$ is calculated using equation [3.35] and gives us a result of 8.7%:

– the product quality ratio $QR_{Al308}(1)$ has a value of 99.34%. This ratio is very close to 1, indicating that very few irreversibilities occur during recycling;

– inequality in the recyclability indicator $RI_{Al308} < RI_{Al308}^{eco}$ defines the limit value RI_{Al308}^{eco} for which it is emergetically suitable to carry out recycling: $RI_{Al308}^{eco} = QR_{Al308}(1) = 99.34\%$, by application of equation [3.53];

– the formation indicator MFI_{Al308} must be lower than its eco-design value $MFI_{Al308}^{eco}(1)$ given by equation [3.53]:

$$\frac{\frac{1}{(1-p_{c,a})(1+a)}MFI^{eco}(1)}{\sum_{i=0}^{1} QR(i)} < 1$$

The value of the formation indicator $MFI_{Al308}^{eco}(1)$ is 2.07. This implies that the unit emergy value mobilized for formation must be less than twice the unit emergy value of the raw material.

If the recyclability indicator *IR*, the additive quantity *a*, and the losses during recycling are constant, the limit formation indicator $MFI_{Al308}^{eco}(\infty)$ (see equation [3.53]) has a value of 1.03. In other words, if the unit emergy value of transformation (formation) is approximately lower than the unit emergy value of the raw material, it is possible to recycle a product an "infinite" number of times (working under the hypotheses above).

3.4. Ratio

Brown and Buranakarn [BRO 03] introduced specific ratios for recycling. These mainly compare the suitability of recycling compared to a situation without recycling. In Figure 3.16, the product at the end of its life is sent to an industrial landfill. The exponent *LF* refers to the landfill. The unit emergy value of the raw material and the refining process are not dependent on use, while the unit emergy value of transformation may depend on the incoming material and is written as em_t^{LF}. Likewise, the unit emergy value of collection and separation is written as em_s^{LF}, and the emergy related to the industrial landfill is written as em_{LF}^{LF}. The notations used in this chapter are retained for the configuration with recycling (see Figure 3.17).

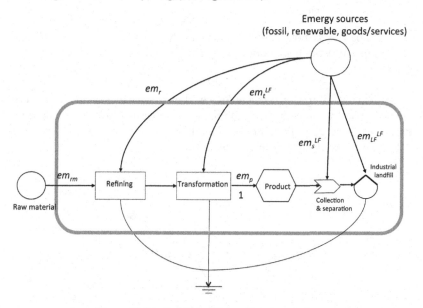

Figure 3.16. *Configuration without recycling*

The landfill ratio LR compares the sum of the unitary emergies for collection and separation and for the industrial landfill for the two configurations; see Figures 3.16 and 3.17:

$$LR = \frac{Em_s^{LF} + Em_{LF}^{LF}}{Em_s + Em_{LF}}$$

Since quantities placed in landfills are not all the same (with or without recycling), the LR ratio makes it possible to gauge the benefit of one solution compared to another. If LR is less than 1, it means that the situation with recycling (and placement in a landfill of subsequent waste) consumes fewer resources than the situation without recycling.

The recycling benefit index[3] (*IBR*) is defined as the fraction between the sum of the unitary emergy of raw material and its refining, and the sum of the unitary emergies of collection and landfill placement in a situation with recycling[4].

$$IBR = \frac{em_p + em_r}{em_c}$$

IBR analysis is not enough to compare two products; a large *IBR* may mean either that the numerator is very large or that the denominator is very small.

In the context of an eco-technological choice, we may seek both to reduce the numerator (a direct process mobilizing fewer resources) and to reduce the denominator (a recycling process mobilizing fewer resources). The *RYB* value is not an indicator to be used alone.

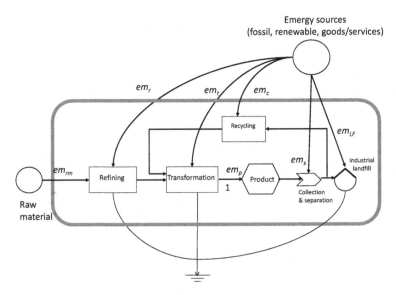

Figure 3.17. *Configuration with recycling*

---

3 The *IBR* index of Brown and Buranakarn [BRO 03] is completely different from the recyclability index *RI* of Jamali-Zghal [JAM 14]: $IMR = \frac{1}{RI}$.

4 100% recycling of a material (even a biodegradable one) is virtually impossible, as part of the raw material is not recycled.

The recycling content ratio i.e. recycle yield ratio (RYB), is defined as the ratio between the sum of the unit emergy values of the raw material, refining and transformation, and the sum of the unit emergy values of collection, separation and recycling:

$$RYB = \frac{em_{rm} + em_t + em_t}{em_s + em_c}$$

RYB analysis is not sufficient to compare two products, for reasons similar to those given above.

3.4.1. Property of a ratio

Current indicators and/or ratios are not all suitable for measuring the suitability of recycling. A dimensionless indicator with the following properties would be useful in the literature:

– a high value for low unit emergy values of material and refining: the choice between two raw materials in ISO use would fall on the material with the lower unitary emergy;

– a high value for low unit emergy values for recycling: the choice between two recycling procedures would fall on the procedure mobilizing fewer resources;

– a high value for low unit emergy values for transformation (formation): the choice between two formation technologies would fall on the procedure using less emergy.

We naturally obtain the triplet:

$$\left(\frac{em_{ref}}{em_{rm} + em_r} ; \frac{em_{ref}}{em_t} ; \frac{em_{ref}}{em_c} \right)$$

from which we can deduce a combination such as:

$$\left(\frac{em_{ref}}{em_{rm} + em_r} \right)^{e1} \left(\frac{em_{ref}}{em_t} \right)^{e2} \left(\frac{em_{ref}}{em_c} \right)^{e3}$$

by introducing the three weightings $e1$, $e2$, and $e3$ (possibly equal to 1).

The reference unit emergy value em_{ref} could be the solar transformity Tr_s or the portion of local renewable emergy $em_{p,R}$. The advantage of a reference unit emergy value that is outside the system being studied is that tendencies are adhered to, while the disadvantage is that this reference is somewhat "artificial".

This indicator would have to be able to cross-reference choices. Two non-exhaustive examples are given below:

– a unit emergy value of raw material that may be higher, but is associated with a lower unitary transformation emergy, should give a high indicator value for ISO use;

– a high-quality raw material (with a high unit emergy value) but with high suitability for recycling might be preferable to a raw material of lower quality but low recyclability for ISO use.

Choices will also have to be reassessed at the time of each recycling cycle.

We would also add that local renewable resources should be favored over ISO products and ISO unit emergy value, following the example of the renewable fraction %R introduced by Wu *et al.* [WU 15].

3.5. Exercise

Calculate the sum of the specific emergy of the raw material and of the refining process $em_{mp} + em_r$, and then the specific transformation emergy em_t, the loss during transformation p_t, the recycled fraction q, and finally the specific recycling emergy em_c. Three materials are proposed: steel, glass and plastic.

3.5.1. *Example application: steel*

	Element	Unit/year	Quantity	Unitary solar emergy seJ/unit	Emergy seJ/year
1	Iron ore	g	4.53E+13	2.83E+09	1.28E+23
2	Natural gas	J	3.17E+17	4.80E+04	1.52E+22
3	Other fuels	J	2.80E+16	6.60E+04	1.85E+21
4	Electricity	J	1.84E+17	1.74E+05	3.20E+22
5	Transport	Ton/mile	7.50E+09	9.65E+11	7.24E+21
6	Work	$	1.58E+09	1.20E+12	1.90E+21
	Annual quantity	g	**4.49E+13**	**To be calculated**	**To be calculated**

Table 3.7. *Emergetic table for steel production using arc-based procedures [BUR 98]*

	Element	Unit/year	Quantity	Unitary solar emergy seJ/unit	Emergy seJ/year
1	Iron ore	g	3.17E+13	2.83E+09	8.97E+22
2	Natural gas	J	3.17E+17	4.80E+04	1.52E+22
3	Other fuels	J	2.80E+16	6.60E+04	1.85E+21
4	Electricity	J	1.84E+17	1.74E+05	3.20E+22
5	Transport	Ton/mile	7.50E+09	9.65E+11	7.24E+21
6	Work	$	1.58E+09	1.20E+12	1.90E+21
7	Recycled steel	g	1.36E+13	2.83E+06	3.85E+22
8	Collection	g	1.3E+13	2.51E+08	3.41E+21
9	Sorting	g	1.36E+13	8.24E+06	1.12E+20
	Annual quantity	g	**4.49E+13**	**To be calculated**	**To be calculated**

Table 3.8. *Emergetic table for steel recycling using arc-based procedures [BUR 98]*

3.5.2. *Example application: glass*

	Element	Unit/year	Quantity	Unitary solar emergy seJ/unit	Emergy seJ/year
1	Silicon sand	g	3.38E+09	1.00E+09	3.38E+18
2	Sand	g	1.31E+08	1.00E+09	1.31E+17
3	Clay	g	1.09E+09	2.00E+09	2.18E+18
4	Other	g	2.18E+08	1.00E+09	2.18E+17
5	Water	J	1.08E+09	4.80E+04	5.18E+13
6	Natural gas	J	8.85E+13	4.80E+04	4.25E+18
7	Electricity	J	1.61E+12	1.74E+05	2.80E+17
8	Transport	Ton/mile	1.19E+06	9.65E+11	1.15E+18
9	Machinery	g	4.08E+07	6.70E+09	2.73E+17
10	Work	$	6.85E+05	1.20E+12	8.22E+17
	Annual quantity	G	**4.14E+09**	**To be calculated**	**To be calculated**

Table 3.9. *Emergetic table for conventional glass production [BUR 98]*

	Element	Unit/year	Quantity	Unitary solar emergy seJ/unit	Emergy seJ/year
2	Sand	g	1.31E+08	1.00E+09	1.31E+17
3	Clay	g	1.09E+09	2.00E+09	2.18E+18
4	Other	g	2.18E+08	1.00E+09	2.18E+17
5	Water	J	1.08E+09	4.80E+04	5.18E+13
6	Natural gas	J	6.65E+13	4.80E+04	3.19E+18
7	Electricity	J	1.21E+12	1.74E+05	2.11E+17
8	Transport	Ton/mile	1.19E+06	9.65E+11	1.15E+18
9	Machinery	g	4.08E+7	6.70E+09	2.73E+17
10	Work	$	6.85E+05	1.20E+12	8.22E+17
11	Glass bottles	g	2.70E+09	1.90E+09	5.13E+18
12	Collection	g	2.70E+09	2.51E+08	6.78E+17
13	Separation	g	2.70E+09	1.32E+07	3.56E+16
	Annual quantity	g	**4.14E+09**	To be calculated	**To be calculated**

Table 3.10. *Emergetic table for glass recycling [BUR 98]*

3.5.3. *Example application: plastic (construction applications)*

	Element	Unit/year	Quantity	Unitary solar emergy seJ/unit	Emergy seJ/year
1	Wood fiber	g	2.67E+12	4.20E+04	1.12E+17
2	Plastic resin	g	7.22E+08	5.27E+09	3.80E+18
3	Electricity	J	1.08E+12	1.74E+05	1.88E+17
4	Transport	Ton/mile	1.87E+05	9.65E+11	1.80E+17
5	Machinery	g	4.84E+05	6.70E+09	3.24E+15
6	Work	$	5.27E+05	1.15E+12	6.06E+17
	Annual quantity	g	**8.50E+08**	To be calculated	**To be calculated**

Table 3.11. *Emergetic table for the conventional production of various plastics [BUR 98]*

	Element	Unit per year	Quantity	Unitary solar emergy seJ/unit	Emergy seJ/year
3	Electricity	J	1.08E+12	1.74E+05	1.88E+17
4	Transport	Ton/mile	1.87E+05	9.65E+11	1.80E+17
5	Machinery	g	4.84E+05	6.70E+09	3.24E+15
6	Work	$	5.27E+05	1.15E+12	6.06E+17
7	Recycled paper	g	2.67E+12	1.45E+05	3.79E+17
8	Recycled plastic	g	7.22E+08	5.27E+09	3.80E+17
9	Collection	g	8.49E+08	2.51E+08	2.13E+18
10	Separation	g	8.49E+08	8.24E+06	7.00E+15
	Annual quantity	g	**8.50E+08**	**To be calculated**	**To be calculated**

Table 3.12. Emergetic table for plastic recycling [BUR 98]

Advanced Concepts in Emergy

4.1. Introduction

Emergy assessment is a concept in progress. As the history of science shows, a great many concepts are controversial at the outset. This is shown more specifically by the history of thermodynamics; until the middle of the 19th Century, the idea that heat was a fluid that flowed to an equilibrium point, like water, was commonly accepted. The formulation of the first law, and the work of Nicolas Carnot [CAR 24], was highly disputed. Even today, the baseline calculation for exergy remains a subject for debate in research journals [SHA 16].

Environmental analyses such as emergy are not without their detractors. There are two major criticisms: the boundaries of the discipline, and the taking into account of history. Certain allocations of inputs (distribution between renewable and non-renewable resources and goods/services) are also the subject of occasional debate.

A part of the scientific community of the International Society of Advancement of Emergy Research (ISAER) has undertaken to establish a database for transformities, and a major axis of ISAER's work consists of proposing gateways through lifecycle analysis [KHA 02]; for an example, see Reza *et al.* [REZ 14].

It is also possible to make so-called theoretical contributions concerning the initial paradigms. This chapter will review ore emergy (see Jamali-Zghal *et al.* [JAM 14]).

4.2. Material emergy: theory review

The methodology of Brown *et al.* [BRO 13] for assessing the unit emergy value of a material (see Chapter 1) presents two problematic issues:

1) It does not fully comply with one of Odum's [ODU 96] ideas designated as the hierarchization of emergy:

> "*In general, rare components are those that require the most work to form and concentrate them. Consequently, they tend to have a high level of transformity. (Burnett, 1981) has shown that the more a component shows a high emergy content, the rarer it is* [ODU 96, p. 21]".

Figure 1.17 shows that silver ore has a low unit emergy value in relation to its concentration.

2) It is not in correlation with material exergy. However, there is a great deal of theoretical research based on thermodynamics used to assess material exergy (see Riekert [RIE 74] and particularly Szargut *et al.* [SZA 02]).

4.2.1. *Transformity: a temporal scale of source exergy*

Transformity (see the definition in section 1.4) and, by extension, unit emergy value, correspond to the history of resources mobilized to obtain their exergetic content; per unit of product in the case of unit emergy value. For Odum, transformity measures the exergetic resources needed to obtain a good, product, or service. From this perspective, transformity corresponds to a scale of exergetic history. In conjunction with this, and analogous to the food chain, a "principle of hierarchization" has been defined as follows: the higher the transformity of a product, the more resources it has required.

The Sun provides abundant, highly diffuse energy. Plants capture this energy via the phenomenon of photosynthesis (the start of the process of concentrating this diffuse solar energy). Herbivores consume the plants. Predators consume the herbivores. This ecosystem is self-regulating, and the "pyramid" of the food chain maintains itself.

Likewise, as described in Chapter 1, the Sun provided energy for an abundance of plant life millions of years ago. Part of the peat generated by this spent time near magma and formed hydrocarbons. Abundant and diffuse energy was concentrated by vegetation, and then part of this was further concentrated to form fossil fuels. This analogy suggests that the more readily available and widespread a quantity is, the lower it is placed on a scale of measurement, and the rarer it is, the higher it should be placed on this same scale. Figure 4.1 illustrates this concept.

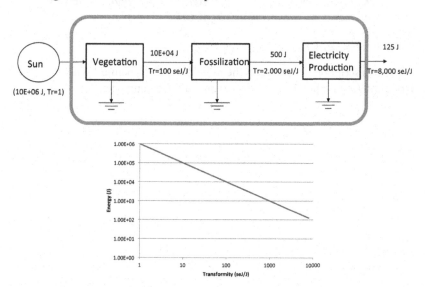

Figure 4.1. *Hierarchization of energy applied to electricity production*

An economist would agree that the system of supply and demand corresponds to this scale (omitting bubble effects). For Odum, transformity must be this value; thus, the higher the transformity of a product, a good, or a service, the more it is the result of a concentration of exergy.

4.2.2. Assessing the UEV of ores using thermodynamic state

The thermodynamic assessment of elements (minerals or metals) carried out by Szargut *et al.* [SZA 05] and extended by Valero *et al.* [VAL 09] allows us to disregard the enrichment factor proposed by Cohen *et al.*

[COH 07] and Brown [BRO 13], on one hand, and to avoid the problem emphasized by Sciubba [SCI 10], on the other.

> *"It is recommended therefore that the Emergy Analysis be not used to assess the global resource consumption caused by anthropogenic activities, because its results are misleading when it comes to estimate the exergy destruction enacted by real industrial transformations [SCI 10]".*

4.2.2.1. *Thermodynamic state approach*

The description of the three theoretical thermodynamic "states" of crust formation (see Figure 4.2) can fit into an emergetic approach.

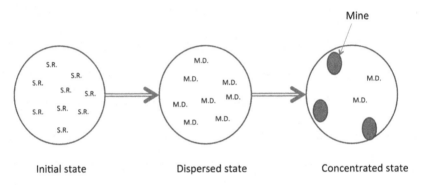

Figure 4.2. *Representation of states for exergetic evaluation of elements*

1) Initial state (reference environment): all substances are dispersed and mixed in a state of equilibrium; the exergy is considered to be zero.

2) Dispersed state (dispersed elements): chemical reactions have taken place; i.e. atoms have formed and elements (minerals and metals) have dispersed in the earth's crust (equidistribution) to reach the concentration $x_{E_{cr}}$. The specific exergy of an element is:

$$ex_E\left(x_{E_{cr}}\right) = ex_{chE} = \Delta g_{fE} + \sum_i y_i \, ex_{ch\,i} \qquad [4.1]$$

with Δg_{f_E} the Gibbs free enthalpy of the element being considered, y_i the molar fraction of the component i and $ex_{ch\,i}$ the specific exergy of this component.

3) Concentrated state (concentrated elements): the dispersed element is locally concentrated in a source, at concentration x_E (directly linked to the enrichment factor of Brown et al. [BRO 13]. The specific exergy of a concentrated element $ex_E(x_E)$ is:

$$ex_E(x_E) = ex_{chE} + ex_{ccE}(x_E) \qquad [4.2]$$

The specific concentration (cc) exergy of the element considered $ex_{ccE}(x_E)$ is obtained by:

$$ex_{ccE}(x_E) = -RT° \left[\ln(\mu_E x_E) + \frac{1-\mu_E x_E}{\mu_E x_E} \ln(1 - \mu_E x_E) \right] \qquad [4.3]$$

The processes described by the reference, dispersed, and concentrated thermodynamic states (yielding sources) are assumed to be of natural origin, as opposed to anthropogenic processes occurring during the exploitation of a source (see Figure 4.3).

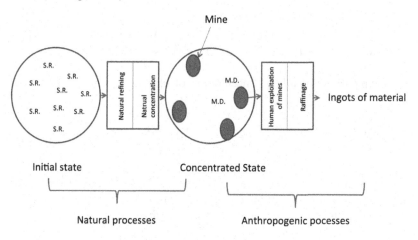

Figure 4.3. *Association of elemental states with exploitation of a source*

The specific exergy in equation [4.2] is obtained considering the processes of reversible thermodynamic states. In order to retain irreversibilities, Valéro *et al.* [VAL 02] have suggested the introduction of factors $k_{ch\,E}$ and $k_{cc\,E}$ to reflect the irreversibilities. The specific exergy of an element is:

$$ex_E(x_E) = k_{ch\,E}\,ex_{chE} + k_{cc\,E}\,ex_{ccE}(x_E) \tag{4.4}$$

Table A.4 gives the values of the factors $k_{ch\,E}$ and $k_{cc\,E}$ as well as the chemical exergy.

Jamali-Zghal [JAM 14] proposed assigning the exergy of the source elements within the source itself (a mixture of earth, for the sake of simplicity, and the element (ore) being exploited). It is thus possible to recover successive thermodynamic states and obtain the specific exergy of the source:

– Dispersed state: natural process:

$$\begin{cases} ex_{Mi}(x_{E_{cr}}) = ex_E^{Mi}(x_{E_{cr}}) = k_{ch\,E}\,ex_{chE} \\ Ex_{Mi}(x_{E_{cr}}) \to 0 \end{cases} \tag{4.5}$$

– Concentrated state: natural process:

$$\begin{cases} ex_{Mi}(x_E^\circ) = ex_E^{Mi}(x_E^\circ) = k_{ch\,E}\,ex_{chE} + k_{cc\,E}\,ex_{ccE}(x_E^\circ) \\ Ex_{Mi}(x_E^\circ) = m_E^\circ\,M_E\,ex_E^{Mi}(x_E^\circ) \end{cases} \tag{4.6}$$

– Exploitation of mine: concentration of element x_E in the diminished source:

$$\begin{cases} ex_{Mi}(x_E) = ex_E^{Mi}(x_E) = k_{ch\,E}\,ex_{chE} + k_{cc\,E}\,ex_{ccE}(x_E) \\ Ex_{Mi}(x_E) = m_E^\circ\,M_E\,ex_E^{Mi}(x_E) \end{cases} \tag{4.7}$$

When the mine is exhausted, the mass of the ore tends toward 0, or, rather, the concentration of the ore tends toward the dispersed concentration. The specific and total exergies of the source are therefore:

$$\begin{cases} \lim_{x_E \to x_{E_{cr}}}\left(ex_{Mi}(x_E)\right) = k_{ch\,E}\,ex_{chE} \\ \lim_{x_E \to x_{E_{cr}}} Ex_{Mi}(x_E) \to 0 \end{cases} \tag{4.8}$$

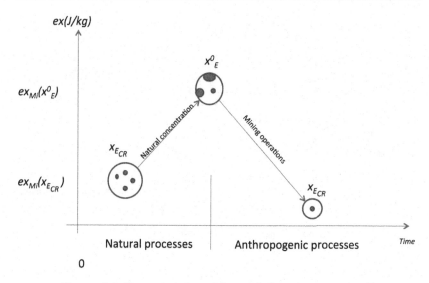

Figure 4.4. *Representation of the evolution of the exergy of an element linked to its concentration*

4.2.2.2. Ore emergies

Jamali-Zghal [JAM 14] proposed the following postulates:

– Postulate 1. Elements (minerals or metals) are by-products of the Earth's formation:

$$Em_E\left(x_{E_{cr}}\right) \overset{\text{def}}{=} Em_{cr} \tag{4.9}$$

By considering the mass of the element at its concentration, we obtain its specific emergy:

$$\begin{cases} m_{E_{cr}} = x_{E_{cr}}\, m_{cr} \\ em_E\left(x_{E_{cr}}\right) = \dfrac{Em_{cr}}{m_{cr}} = \dfrac{em_{cr}}{x_{E_{cr}}} \end{cases} \tag{4.10}$$

– Postulate 2. The origin of the exergy of the different thermodynamic states has a single initial source (see Figure 4.5):

$$Tr_I \overset{\text{def}}{=} \frac{Em_E(x_{E_{cr}})}{Ex_E(x_{E_{cr}})} \overset{\text{def}}{=} \frac{Em_{Mi}(x_E^\circ)}{Ex_{Mi}(x_E^\circ)} \tag{4.11}$$

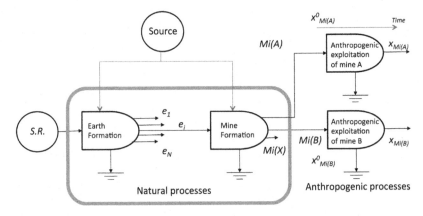

Figure 4.5. *Association of elemental formation*
states with an emergetic approach

The specific emergy of the mine before the start of its exploitation is given by:

$$em_{Mi}(x_E^{\circ}) \quad = em_E(x_{Ecr}) \frac{ex_{Mi}(x_E^{\circ})}{ex_E(x_{Ecr})}$$

$$= \frac{em_{cr}}{x_{Ecr}} \left[\frac{k_{ch\,E}\,ex_{chE} + k_{cc\,E}\,ex_{ccE}(x_E^{\circ})}{k_{ch\,E}\,ex_{chE}} \right] \qquad [4.12]$$

$$= \frac{em_{cr}}{x_{Ecr}} \left[1 + \frac{k_{cc\,E}\,ex_{ccE}(x_E^{\circ})}{k_{ch\,E}\,ex_{chE}} \right]$$

The total emergy of a mine before the start of its exploitation is:

$$Em_{Mi}(x_E^{\circ}) = m_E^{\circ}\,em_{Mi}(x_E^{\circ}) \qquad [4.13]$$

$$\lim_{x_E \to x_{Ecr}} em_{Mi}(x_E) = em_E(x_{Ecr})$$

In the theoretical case of a planet composed only of a single element, i.e. $x_{Ecr} \to 1$, we have:

$$\lim_{x_{Ecr} \to 1} em_{Mi}(x_E) = em_{cr}$$

– Postulate 3. The transformity of a mine being exploited does not vary, no matter the concentration of the ore:

$$Tr_I \stackrel{\text{def}}{=} \frac{Em_{Mi}(x_E^\circ)}{Ex_{Mi}(x_E^\circ)} \stackrel{\text{def}}{=} \frac{Em_{Mi}(x_E)}{Ex_{Mi}(x_E)} \qquad [4.14]$$

$$
\begin{aligned}
em_{Mi}(x_E) &= em_{Mi}(x_E^\circ) \frac{ex_{Mi}(x_E)}{ex_{Mi}(x_E^\circ)} \\
&= \frac{em_{cr}}{x_{Mcr}} \left[\frac{k_{ch\,E}\,ex_{chE} + k_{cc\,E}\,ex_{ccE}(x_E)}{k_{ch\,E}\,ex_{chE}} \right] \qquad [4.15] \\
&= \frac{em_{cr}}{x_{Mcr}} \left[1 + \frac{k_{cc\,E}\,ex_{ccE}(x_E)}{k_{ch\,E}\,ex_{chE}} \right]
\end{aligned}
$$

$$Em_{Mi}(x_E) = m_E\,em_{Mi}(x_E) \qquad [4.16]$$

$$\lim_{x_E \to x_{Ecr}} em_{Mi}(x_E) = em_{Mi}(x_{Ecr})$$

$$\lim_{x_E \to x_{Ecr}} Em_{Mi}(x_E) = 0$$

– Postulate 4. The total emergy of the mine Em_{Mi} is equal to the total emergy of its ore Em_E:

$$Em_E(x_E) = Em_{Mi}(x_E) \qquad [4.17]$$

$$m_E = x_E\,m_{Mi}$$

$$em_E(x_E) = x_E\,em_{Mi}(x_E) = \frac{x_E}{x_{Ecr}} \left[1 + \frac{k_{cc\,E}\,ex_{ccE}(x_E)}{k_{ch\,E}\,ex_{chE}} \right] em_{cr} \qquad [4.18]$$

$$\lim_{x_E \to x_{Ecr}} em_{Mi}(x_E) = em_{cr}$$

The emergy required for a backfill is equal to the mass extracted during the exploitation phase of the mine times the specific emergy of the terrestrial crust (see Figure 4.6):

$$Em_{rem} = m_{rem}\,em_{cr}$$

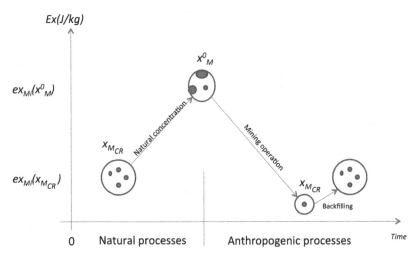

Figure 4.6. *Exergetic representation of the
backfilling of a mine*

The model described by equations [4.9]–[4.18] requires us to determine the transformity of the initial source introduced in Postulate 2, written as Tr_I. Jamali-Zghal [JAM 14] proposed choosing the transformity of the geobiosphere. A geobiosphere model was shown in Chapter 1; here, sources (Sun, lunar gravity for tides, natural radioactivity, the heat of the Earth's core) are used to determine the annual flow of exergy received by the geobiosphere taken as a whole, which corresponds roughly to solar energy (3.6E+24 J). The transformity is therefore the ratio between the total emergy (15.2E+24 seJ/year) and the exergy of the biosphere:

$Tr_I=4.2 \ seJ/J.$

Table 4.1 lists the elements selected in Chapter 1 and reiterates the specific emergy calculated by the enrichment factor method, and then gives the specific emergy value in the dispersed state $x_{E_{cr}}$ (or after a mine has been exploited). The last column gives the specific emergy up to a minimum threshold of profitability (MTP). It is advisable, then, to take into account the concentration of an ore during its extraction phase in order to establish its transformity during subsequent applications (as a source).

	UEV seJ/g Enrichment factor	UEV seJ/g Jamali-Zghal by-product	UEV seJ/g Jamali-Zghal by-product
Earth's crust	1.35E+08	Dispersed	MTP
Aluminum	2.86E+08	1.23E+03	2.25E+07
Iron	1.93E+09	2.81E+03	4.68E+07
Titanium	4.49E+09	3.28E+04	7.62E+08
Manganese	5.61E+10	1.64E+05	1.92E+10
Chromium	8.86E+11	2.81E+06	1.15E+10
Nickel	6.06E+10	4.92E+06	3.28E+09
Copper	2.70E+10	3.94E+06	1.30E+09
Lead	2.70E+11	4.92E+06	2.66E+09
Silver	2.16E+07	1.97E+09	3.70E+13
Mercury	1.68E+12	1.23E+09	7.23E+12
Gold	1.12E+14	5.47E+10	8.35E+18

Table 4.1. *Comparison of UEV according to approach and ore*

In Figure 4.7, the specific emergy of the elements proposed is compliant with the hierarchization of energies as put forth by Odum [ODU 96].

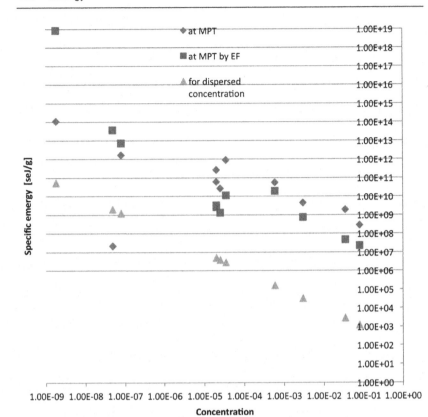

Figure 4.7. *Specific emergy of elements at different concentrations (see Appendix, Table A.1). For a color version of the figure, see www.iste.co.uk/lecorre/emergy.zip*

4.2.2.3. Principal consequences

– Considering the formation of elements as by-products during the formation of the Earth poses the problem of by-products. In the use of emergetic tables and by applying Rule 4 (section 1.1), it would no longer be the sum of the "materials" but rather the maximum of these terms that would need to be taken into account for the elements considered. However, the transformity of a material is comprised of two parts; one of them natural (the maximum of which must be used) and the other anthropogenic (for which the sum must be calculated). Each material would therefore no longer be shown as a line in an emergy table, but rather as two lines; the first representing the natural part and the second the anthropogenic part.

– Taking into account the exhaustion of the mine introduces two problems hindering easy implementation:

- it is necessary to take concentration into account during exploitation. A mine does not have a constant concentration, and ore does not have the same specific emergy, regardless of the emergetic investment made by exploitation. There is a temporal variation in emergy;

- ore emergies vary according to the location of the mine. There is no reason why every mine in the world should be at the same concentration.

These difficulties of implementation explain why Jamali-Zghal's model [JAM 14] is seldom used in applications. However, it is closer to Odum's original concept [ODU 96], in particular in its hierarchization of elements.

– A major objection posed by Valero [VAL 08, p. 8] is worded as follows:

> "No matter how much solar energy is received from the Sun, the quantity of gold or iron for instance on Earth, will not change. Consequently, the rigorousness of the transformities for mineral resources assessment is doubtful."

Jamali-Zghal [JAM 14] proposed a response by combining Szargut's initial approach and the introduction of irreversibilities put forth by Valero *et al.* [VAL 02] in the calculation of specific emergy. Yet, the choice of initial transformity $Tr_I=4.2$ seJ/J is crucial, as it positions the materials in comparison to fossil resources. However, Figure 4.7 shows that the value chosen by Jamali-Zghal [JAM 14] is consistent with the previous values of Brown [BRO 13].

If we look at emergy as the history of the exergy mobilized in order to obtain a product, good, or service, the model of Jamali-Zghal [JAM 14] is consistent. The major question of knowing what this source was doing at the Big Bang goes far beyond the purpose of emergetic analysis.

4.2.3. *Example of implementation: copper production in Australia*

The USGS database reports on annual ore production, including that of Australia. The example used here is that of copper. Figure 4.8 shows extraction on the left-hand ordinate and concentration on the right-hand ordinate. Copper concentration decreases as years of exploitation increase, with high-concentration deposits already having been exploited.

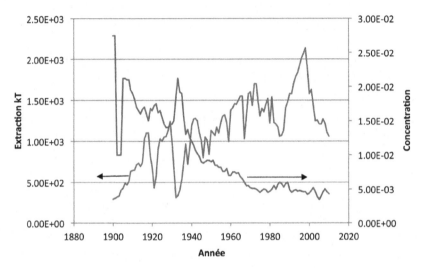

Figure 4.8. *Average concentration and tonnage extracted from Australian mines. For a color version of the figure, see www.iste.co.uk/lecorre/emergy.zip*

Using Table A.4 in the Appendix and applying equation [4.7], we can trace the molar exergy of the mines' production (depending on the concentration of ore according to the year considered) (see Figure 4.9).

Equation [4.15] can be used to calculate the unit emergy value of the rock taken from a copper mine (see Figure 4.10). Figure 4.11 shows that the lower the concentration, the higher the unit emergy value of the extracted rock, as expected by Odum [ODU 96, p. 121].

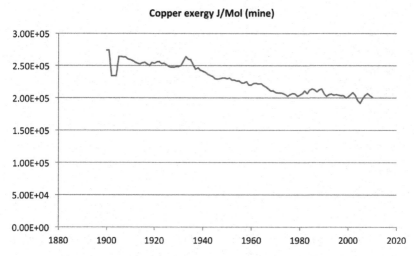

Figure 4.9. *Exergy of copper taken from Australian mines*

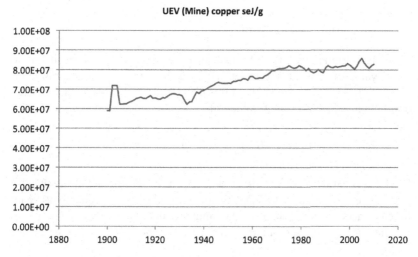

Figure 4.10. *Unitary emergetic value of copper rock taken from Australian mines*

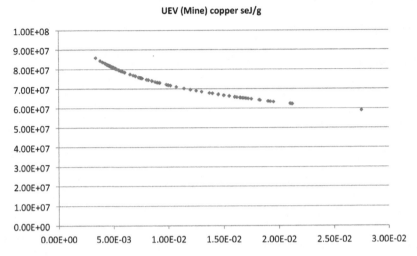

Figure 4.11. *Unitary emergetic value of copper rock according to concentration*

4.2.4. Hubbert's peak and emergy

Hubbert starts from the physical idea that a mine or a deposit has an initial value of a quantity (ore or fuel) and that anthropogenic exploitation does nothing but push the value of this quantity toward exhaustion. We use a Gaussian function as an approximation:

$$HF(t) = \widehat{FH} \exp\left(-0.5\left(\frac{t-t_{peak}}{s}\right)^2\right) \qquad [4.19]$$

If we consider a peak, written as t_{peak}, in 2040, a maximum value $\widehat{HF}$ taken to be 100, and a curve shape factor s considered to be equal to 40 years, we can trace the rough shape of the Hubbert curve (see Figure 4.12). Let us assume that the concentration of an ore is proportional to the remaining tonnage:

$$x(t) = \frac{m^0 - m(t)}{m^0} x^0 \qquad [4.20]$$

Knowing the initial state of a mine or deposit (characterized by its initial mass and concentration), we can use Hubbert's model to deduce the remaining quantity depending on exploitation, and to deduce from this a temporal variation of concentration. An example is given in Figure 4.12.

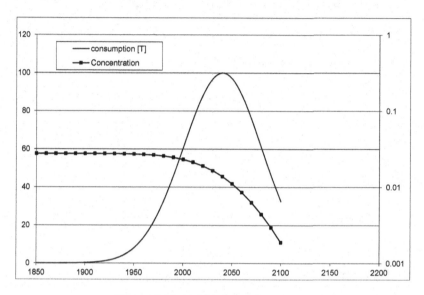

Figure 4.12. *Hubbert peak for copper*
and modeled concentration

It is then possible to calculate the unit emergy value of the rock, which is an increasing function for a reduction in concentration (see Figure 4.13).

This tendency is of particular interest as it gives more sense to the fact that the unit emergy value of a recycled material is an increasing function with recycling. In reality, the more a mine is exhausted, the more difficult it becomes to extract ore from it, and the augmentation of the unit emergy value of recycled materials is put in competition with this exhaustion effect. Thus emergy regains its full meaning as a value connected to sustainable development as defined by the Bruntland Commission [WCE 87].

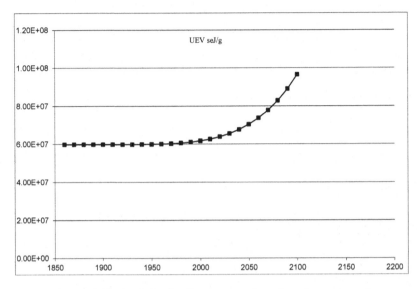

Figure 4.13. *Growth of unitary emergetic value of copper rock for decreasing concentration due to exploitation*

4.3. Conclusion

This book explains the fundamental principles of emergy (see Chapter 1). The application of calculation rules and the paradigm of evaluating the emergy of the geobiosphere are discussed in detail. In Chapter 2 numerous implicit-time examples are given, which are directly linked to the concept of sustainable development. Following the history/pathway of the material, as in the Lagrangian approach, emergy analysis is applied to recycling; see Chapter 3. Finally, Chapter 4 describes the results of recent research.

The impermanence of unit emergy values and the interdependence of a product, good, or service with its environment are two concepts intrinsic to emergy analysis. These two concepts, which are counterintuitive and thus debatable given their basis in hypothesis, are inherent in the concept of sustainable development.

In the Introduction, a distinction was made between resource-oriented methods and product-oriented methods, as is commonly done in the literature. However, they present the same fundamental difficulties, though a norm appears to propose a more specific context, particularly in the case of lifecycle analysis. Discrete-time equations put forth in the context of recycling should also appear in LCA.

When I accepted Mr Feidt's invitation, I didn't quite understand the amount of work that goes into writing a book – but I left my comfort zone and accepted the challenge. I hope readers have found the information in this book useful, interesting and thought-provoking.

"They did not know it was impossible, so they did it."

Mark Twain

Appendix

Unitary Emergetic Value of Ores

One	Chemical formula	Molar mass	Concentration	Limit quality	Enrichment factor
Aluminum	Al2O3	102	8.00E-02	1.70E-01	2.13E+00
Antimony	Sb2S3	339.7	2.00E-07	1.00E-01	5.00E+05
Arsenic	FeAsS	162.8	1.50E-06	2.00E-02	1.33E+04
Barium	BaSO4	–	5.50E-04	5.80E-02	1.05E+02
Beryllium		537.6	3.00E-06	4.00E-02	1.33E+04
Bismuth	Bi2S3	514.2	1.27E-07	5.00E-02	3.94E+05
Cadmium	CdS	144.5	9.80E-08	4.00E-03	4.08E+04
Cesium	CsCl	–	3.70E-06	2.00E-01	5.41E+04
Chromium	FeCr2O4	223.8	3.50E-05	2.30E-01	6.57E+03
Cobalt	CoS2	123.1	1.00E-05	2.00E-03	2.00E+02
Copper	CuFeS2	183.5	2.50E-05	5.00E-03	2.00E+02
Fluoride	CaF2	78.1	6.50E-04	1.00E-03	1.54E+00
Gallium	Ga(OH)3	–	1.70E-05	1.00E-04	5.88E+00
Germanium	GeO2	–	1.60E-06	6.00E-02	3.75E+04
Gold	Au	197	1.80E-09	1.50E-05	8.33E+03
Hafnium	HfO2	–	5.80E-06	4.00E-04	6.90E+01
Indium	InS	146.9	5.00E-08	1.00E-04	2.00E+03
Iron	Fe2O3	159.7	3.50E-02	5.00E-01	1.43E+01

Lead	PbS	239.3	2.00E-05	4.00E-02	2.00E+03
Lithium	LiAlSi2O6	186.1	2.00E-05	3.00E-02	1.50E+03
Magnesium	MgCl2	95.2	1.33E-02	1.00E-03	7.52E-02
Manganese	MnO2	86.9	6.00E-04	2.50E-01	4.17E+02
Mercury	HgS	232.7	8.00E-08	1.00E-03	1.25E+04
Molybdenum	MoS2	160.1	1.50E-06	3.00E-03	2.00E+03
Nickel	NiS	90.7	2.00E-05	9.00E-03	4.50E+02
Niobium	Nb2O5	–	2.50E-05	5.00E-03	2.00E+02
Phosphorus	P2O5	142	7.00E-04	1.48E-01	2.11E+02
Platinum	PtS	–	1.00E-08	1.95E-05	1.95E+03
Potassium	KCl	74.56	2.80E-02	1.50E-01	5.36E+00
Rhenium	ReS2	250.3	4.00E-10	3.00E-03	7.50E+06
Selenium	SeO2	–	5.00E-05	2.50E-06	5.00E-02
Silicon	SiO2	60.1	3.08E-01	4.00E-01	1.30E+00
Silver	Ag2S	247.8	5.00E-08	1.00E-04	2.00E+03
Sodium	NaCl	58.4	2.90E-02	2.00E-01	6.90E+00
Tantalum	Ta2O5	441.1	2.20E-06	1.00E-03	4.55E+02
Tellurium	TeO2	–	1.00E-09	1.00E-06	1.00E+03
Tin	SnO2	150.7	5.50E-06	4.00E-03	7.27E+02
Titanium	FeTiO3	151.7	3.00E-03	1.00E-01	3.33E+01
Tungsten	CaWO4	288	2.00E-06	6.00E-03	3.00E+03
Vanadium	V2O5	182	6.00E-05	6.00E-03	1.00E+02
Zinc	ZnS	97.4	7.10E-05	3.50E-02	4.93E+02
Zirconium	ZrSiO4	183.3	1.90E-04	2.00E-02	1.05E+02

Table A.1. *Principal ores*

One	Chemical formula	UEV in seJ/g with EF	UEV at seJ/g in dispersed state	UEV in seJ/g at MTP
Aluminum	Al2O3	2.84E+08	1.23E+03	2.25E+07
Antimony	Sb2S3	6.75E+13	4.92E+08	1.86E+10
Arsenic	FeAsS	1.80E+12	6.56E+07	8.71E+09
Barium	BaSO4	1.42E+10	1.79E+05	–
Beryllium	6SiO2:Al2O3: 3BeO	1.80E+12	3.28E+07	1.32E+12
Bismuth	Bi2S3	5.31E+13	7.75E+08	1.01E+11
Cadmium	CdS	5.51E+12	1.00E+09	2.86E+12
Cesium	CsCl	7.30E+12	2.66E+07	–
Chromium	FeCr2O4	8.87E+11	2.81E+06	1.15E+10
Cobalt	CoS2	2.70E+10	9.84E+06	1.18E+10
Copper	CuFeS2	2.70E+10	3.94E+06	1.30E+09
Fluoride	CaF2	2.03E+08	1.51E+05	5.60E+06
Gallium	Ga(OH)3	7.97E+08	5.79E+06	–
Germanium	GeO2	5.06E+12	6.15E+07	–
Gold	Au	1.12E+12	5.47E+10	8.35E+18
Hafnium	HfO2	9.32E+09	1.70E+07	–
Indium	InS	2.70E+11	1.97E+09	–
Iron	Fe2O3	1.93E+09	2.81E+03	4.68E+07
Lead	PbS	2.70E+11	4.92E+06	2.66E+09
Lithium	LiAlSi2O6	2.03E+11	4.92E+06	4.91E+11
Magnesium	MgCl2	1.35E+07	7.40E+03	3.02E+05
Manganese	MnO2	5.63E+10	1.64E+05	1.92E+10
Mercury	HgS	1.69E+12	1.23E+09	7.23E+12
Molybdenum	MoS2	3.06E+11	6.56E+07	6.81E+11
Nickel	NiS	6.08E+10	4.92E+06	3.28E+09
Niobium	Nb2O5	2.54E+10	3.94E+06	–
Phosphorus	P2O5	2.85E+10	1.41E+05	2.27E+08
Platinum	PtS	2.63E+11	9.84E+09	–
Potassium	KCl	7.79E+12	3.51E+03	2.59E+07
Rhenium	ReS2	4.05E+14	2.46E+11	1.16E+15

Selenium	SeO2	3.38E+09	1.97E+06	–
Silicon	SiO2	1.76E+08	3.19E+02	5.17E+05
Silver	Ag2S	2.70E+11	1.97E+09	3.70E+13
Sodium	NaCl	9.32E+08	3.39E+03	3.60E+07
Tantalum	Ta2O5	3.38E+10	4.47E+07	1.86E+14
Tellurium	TeO2	1.35E+10	9.84E+10	–
Tin	SnO2	8.59E+10	1.79E+07	1.02E+13
Titanium	FeTiO3	4.50E+09	3.28E+04	7.62E+08
Tungsten	CaWO4	4.05E+11	4.92E+07	4.20E+13
Vanadium	V2O5	5.79E+09	1.64E+06	3.32E+11
Zinc	ZnS	6.66E+10	1.39E+06	1.81E+08
Zirconium	ZrSiO4	1.42E+10	5.18E+05	1.50E+12

Table A.2. *Unitary emergetic value (UEV) for various components; 1st column using the enrichment factor method; 2nd column using the Jamali–Zghal method at average terrestrial concentration and then at the minimum threshold of profitability*

FRANCE		Quantity	Unit	UEV	Emergy
					E20 seJ/year

RENEWABLE ENERGY:

		Quantity	Unit	UEV	Emergy
1	Sun	1.40E+21	J	1.00E+00	14.1
2	Crustal heat	Unavail.	J	2.03E+04	Unavail.
3	Tide	4.60E+18	J	7.24E+04	3312.3
4	Wind	5.80E+18	J	1.58E+03	91.8
5	Water	–	J	Variable	198.2
6	Waves	Unavail.	J	2.22E+04	Unavail.

INTERNAL TRANSFORMATIONS (ECONOMY):

		Quantity	Unit	UEV	Emergy
7	Agriculture	1.20E+18	J	Variable	1648.2
8	Cereals	1.30E+17	J	variable	3510.9
9	Fishing	1.90E+15	J	8.40E+06	161.3
10	Wood	1.80E+17	J	Variable	562.3

11	Forestry	1.70E+17	J	Variable	77.9
12	Water industry	1.60E+17	J	2.40E+05	383.6
13	Hydroelectricity	2.30E+17	J	2.80E+05	629.1
14	Electricity (other)	1.70E+18	J	2.90E+05	4745.9

NON-RENEWABLE LOCAL EXTRACTION:

15	Forests	0.00E+00	J	3.80E+04	0
16	Fish	3.30E+14	J	8.40E+06	27.6
17	Water	0.00E+00	J	2.80E+05	0
18	Topsoil loss, organic matter	2.10E+16	J	Variable	62.9
19	Coal	0.00E+00	J	8.20E+04	0
20	Natural gas	5.60E+16	J	1.70E+05	96
21	Oil	6.30E+16	J	1.50E+05	93.7
22	Minerals	1.70E+13	g	Variable	1484
23	Metals	1.60E+11	g	Variable	4.1

IMPORTED:

24	Oil	–	Mixed	Variable	12852.3
25	Metals	–	Mixed	Variable	7058.9
26	Minerals	–	Mixed	Variable	531.8
27	Food and agricultural products	–	Mixed	Variable	713.8
28	Cereals, meat, fish	–	Mixed	Variable	842.2
29	Plastics, etc.	–	g	Variable	1063.2
30	Chemicals	–	Mixed	Variable	1053
31	Finished products	–	Mixed	Variable	878.5
32	Machinery & equipment, transport	2.20E+11	$	2.70E+12	3332.3
33	Misc. goods	1.30E+11	$	2.70E+12	3404.7
34	Electricity (other)	3.80E+16	J	2.90E+05	110
35	Importation services	6.90E+11	$	2.70E+12	18679.3

EXPORTS:

36	Fuel	–	Mixed	Variable	2167.1
37	Metals	–	Mixed	Variable	3562.2
38	Minerals	–	Mixed	Variable	299.1
39	Agricultural products and food	–	Mixed	Variable	1409.9
40	Cereals, meat, fish	–	Mixed	Variable	822.5
41	Plastics, etc.	–	Mixed	Variable	1187.7
42	Chemicals	–	Mixed	Variable	566.5
43	Finished products	–	Mixed	Variable	644.9
44	Mach. & equip. transp.	2.30E+11	$	1.90E+12	2971.7
45	Misc. goods	1.00E+11	$	1.90E+12	1643.8
46	Electricity (other)	2.10E+17	J	2.90E+05	604.2
47	Importation services	5.90E+11	$	1.90E+12	11514.1
48	Tourism	6.70E+10	$	1.90E+12	1295.7

Table A.3. *Economic data for France (2008)*

Element	Chemical formula	Irreversibility *kcc*	Chemical irreversibility *kch*	Chemical exergy kJ/mol
Aluminum	Al2O3	3.96E+02	1	3.49E+01
Antimony	Sb2S3	2.84E+01	10	2.52E+03
Arsenic	FeAsS	7.99E+01	10	1.43E+03
Barium	BaSO4	NA	1	–
Beryllium	6SiO2:Al2O3: 3BeO	1.12E+02	1	6.58E+01
Bismuth	Bi2S3	8.98E+01	10	2.23E+03
Cadmium	CdS	8.04E+02	10	7.44E+02
Cesium	CsCl	NA	1	–
Chromium	FeCr2O4	3.67E+01	1	1.95E+02
Cobalt	CoS2	1.26E+03	10	1.39E+03
Copper	CuFeS2	3.86E+02	10	1.53E+03
Fluoride	CaF2	1.70E+00	1	5.03E+01
Gallium	Ga(OH)3	–	1	–
Germanium	GeO2	–	1	–

Gold	Au	4.09E+05	1	5.99E+01
Hafnium	HfO2	–	1	–
Indium	InS	-4.39E+01	10	9.13E+02
Iron	Fe2O3	4.40E+01	1	1.64E+01
Lead	PbS	2.12E+02	10	7.41E+02
Lithium	LiAlSi2O6	1.58E+02	1	2.87E+01
Magnesium	MgCl2	1.00E+00	1	1.61E+02
Manganese	MnO2	2.84E+02	1	3.49E+01
Mercury	HgS	1.71E+03	10	6.79E+02
Molybdenum	MoS2	9.47E+02	1	1.72E+03
Nickel	NiS	3.37E+02	10	7.66E+02
Niobium	Nb2O5	NA	1	–
Phosphorus	P2O5	4.39E+01	1	3.59E+02
Platinum	PtS	NA	1	–
Potassium	KCl	3.86E+01	1	1.96E+01
Rhenium	ReS2	1.94E+03	10	1.62E+03
Selenium	SeO2	NA	1	–
Silicon	SiO2	1.89E+00	1	8.20E-01
Silver	Ag2S	7.05E+03	10	7.06E+02
Sodium	NaCl	3.81E+01	1	1.44E+01
Tantalum	Ta2O5	1.25E+04	1	4.56E+01
Tellurium	TeO2	NA	1	–
Tin	SnO2	1.49E+03	1	4.26E+01
Titanium	FeTiO3	3.48E+02	1	1.30E+02
Tungsten	CaWO4	3.10E+03	1	7.21E+01
Vanadium	V2O5	5.72E+02	1	3.23E+01
Zinc	ZnS	6.28E+01	10	7.44E+02
Zirconium	ZrSiO4	7.74E+03	1	3.08E+01

Table A.4. *Concentration and chemical irreversibility factor introduced by Valero et al. [VAL 02], and chemical exergy of substances*

Bibliography

[ADE 10] ADEME, Emission Factor Guide, Ministry of Environment and Energy, France, 2010.

[ALV 12] ALVES DA CRUZ R., OLLER DO NASCIMENTO C., "Emergy analysis of oil production from microalgae", *Biomass and Bioenergy*, vol. 47, pp. 418–425, 2012.

[AND 15] ANDRIC I., JAMALI–ZGHAL N., SANTARELLI M. *et al.*, "Environmental performance assessment of retrofitting existing coal fired power plants to co-firing with biomass: carbon footprint and emergy approach", *Journal of Cleaner Production*, vol. 103, pp. 13–27, 2015.

[BAR 04] BARGIGLI S., RAUGEI M., ULGIATI S., "Comparison of thermodynamic and environmental indexes of natural gas, syngas and hydrogen production processes", *Energy*, vol. 29, pp. 2145–2159, 2004.

[BAS 96] BASTIANONI S., MARCHETTINI N., "Ethanol production from biomass: analysis of process efficiency and sustainability", *Biomass Bioenergy*, vol. 5, no. 11, pp. 411–418, 1996.

[BEE 99] BEERLING D., "Quantitative estimates of changes in marine and terrestrial primary productivity over the past 300 million years", *Biological Sciences*, vol. 266, pp. 1821–1827, 1999.

[BER 09] BERTRAM M., MARTCHEK K., ROMBACH G., "Material flow analysis in the aluminum industry", *Journal of Industrial Ecology*, vol. 13, no. 5, pp. 650–654, 2009.

[BRO 03] BROWN M., BURANAKARN V., "Emergy indices and ratios for sustainable material cycles and recycle options", *Resources, Conservation and Recycling*, vol. 38, no. 1, pp. 1–22, 2003.

[BRO 10] BROWN M., ULGIATI S., "Updated evaluation of exergy and emergy driving the geobiosphere: a review and refinement of the emergy baseline", *Ecological Modelling*, vol. 221, no. 20, pp. 2501–2508, 2010.

[BRO 11] BROWN M., PROTANO G., ULGIATI S., "Assessing geobiosphere work of generating global reserves of coal, crude oil, and natural gas", *Ecological Modelling*, vol. 222, no. 3, pp. 879–887, 2011.

[BRO 13] BROWN M., "Material Cycles and Energy Hierarchy", available at: http://www.cep.ees.ufl.edu/emergy/resources/presentations.shtml, 2013.

[BS 06] BS EN ISO 1404, Environmental Management – Life Cycle Assessment – Requirements and Guidelines, European Committee for Standardization, Brussels, Belgium, 2006.

[BUR 81] BURNETT M., "A methodology for assessing net energy and abundance of energy resources", *Energy and Ecological Modelling*, pp. 703–710, 1981.

[BUR 98] BURANAKARN V., Evaluation of Recycling and Reuse of Building Materials Using the Emergy Analysis Method, thesis, University of Florida, Gainesville, FL, 1998.

[CAR 24] CARNOT S., *Réflexions sur la puissance motrice du feu*, Blanchard, Paris, 1824.

[CAS 04] CASTRO M., REMMERSWAAL J., REUTER, M. *et al.*, "A thermodynamic approach to the compatibility of materials combinations for recycling", *Resources, Conservation and Recycling*, vol. 43, no. 1, pp. 1–19, 2004.

[CAS 07] CASTRO M., REMMERSWAAL J., "Exergy losses during recycling and the resource efficiency of product systems", *Resources, Conservation and Recycling*, vol. 38, no. 1, pp. 1–22, 2007.

[COH 07] COHEN M., SWEENY S., BROWN M., "Computing the unit emergy value of crustal elements", *Proceedings of the Fourth Biennal Emergy Reasearch Conference*, Gainesville, FL, pp. 16.1–16.12, 2007.

[COH 12] COHEN M., SWEENEY S., KING D. *et al.*, *Environmental Accounting of National Economic Systems*, United Nations Environment Programme, available at: http://www.unep.org/dewa/Portals/67/pdf/EANE_Report_lowres.pdf, 2012.

[COP 09] COPPOLA F., BASTIANONI S., ØSTERGÅRD H., "Sustainability of bioethanol production from wheat with recycled residues as evaluated by Emergy assessment", *Biomass Bioenergy*, vol. 33, pp. 1626–1642, 2009.

[CRA 09] CRAWFORD, R., "Life cycle energy and greenhouse emissions analysis of wind turbines and the effect of size on energy yield", *Renewable and Sustainable Energy Reviews*, vol. 13, pp. 2653–2660, 2009.

[DEL 13] DELRUE, F., Les différentes techniques de récolte de micro-algues: aspects techniques, économiques et environnementaux, available at: http://mio.pytheas. univ-amu.fr/IMG/pdf/20131022_02_presa_CEA.pdf, 2013.

[DOE 15] DOE, Weather data, US Department of Energy, available at: http://apps1. eere.energy.gov/buildings/energyplus/weatherdata_about.cfm, 2015.

[DUK 03] DUKES, J., "Burning buried sunshine: human consumption of ancient solar energy", *Climatic Change*, vol. 61, pp. 31–44, 2003.

[EPA 05] EPA, Emission Facts, Average CO_2 Emissions Resulting from Gasoline and Diesel Fuel, Environmental Protection Agency of the USA, 2005.

[FIN 97] FINNVEDEN G., ÖSTLUND P., "Exergies of natural resources in life-cycle assessment and other applications", *Energy*, vol. 22, no. 9, pp. 923–931, 1997.

[HER 06] HERMANN W., "Quantifying global exergy resources", *Energy*, vol. 31, no 12, pp. 1685–1702, 2006.

[IGN 07] IGNATENKO O., VAN SCHAIK A., REUTER M., "Exergy as a tool for evaluation of the resource efficiency of recycling systems", *Minerals Engineering*, vol. 20, no. 9, pp. 862–874, 2007.

[ISA 15] ISAER, The Emergy Database, available at: www.emergydatabase.org, 2015.

[JAM 13] JAMALI–ZGHAL N., AMPONSAH N., LACARRIERE B. *et al*, "Carbon footprint and emergy combinatin for eco-environmental assessment of cleaner heat production", *Journal of Cleaner Production*, vol. 47, pp. 446–456, 2013.

[JAM 14a] JAMALI–ZGHAL N., Environmental assessment tools for sustainable resource management, Thesis, Ecole des Mines de Nantes, France, 2014.

[JAM 14b] JAMALI–ZGHAL N., LE CORRE O., LACARRIERE B., "Mineral resource assessment: compliance between emergy and exergy respecting Odum's hierarchy concept", *Ecological Modelling*, vol. 272, pp. 208–219, 2014.

[KHA 02] KHALIFA K., "Analyse du cycle de vie: Méthodes d'évaluation des impacts", *Techniques de l'ingénieur*, p. 22, 2002.

[LEC 12] LE CORRE O., TRUFFET L., "Exact computation of emergy based on a mathematical reinterpretation of the rules of emergy algebra", *Ecological Modelling*, vol. 230, pp. 101–113.

[LI 10] LI L., LU H., CAMPBELL D. *et al.*, "Emergy algebra: improving matrix methods for calculating transformities", *Ecological Modelling*, vol. 221, no. 3, pp. 411–422, 2010.

[MEN 06] MENG S., TAO Z., WU B. *et al.*, "Study on burning clinker using the matrix bonding component in waste concrete as a raw material component", *Cement Engineering*, vol. 1, pp. 1–5, 2006

[NEA 15] NEAD, National Environmental Accounting Database, available at: http://www.cep.ees.ufl.edu/nead/index.php, 2015.

[NIM 15] NIMMANTERDWONG P., CHALERMSINSUWAN B., PIUMSOMBOON P., "Emergy evaluation of biofuels production in Thailand from different feedstocks", *Ecological Engineering*, vol. 74, pp. 423–437, 2015.

[ODU 95] ODUM H., PETERSON N., "Simulation and evaluation with energy systems blocks", *Ecological Modelling*, vol. 93, nos. 1–3, pp. 155–173, 1995.

[ODU 96] ODUM H., *Environmental Accouting: Emergy and Decision Making*, John Wiley, New York, 2015.

[ODU 00a] ODUM H., *Handbook of Emergy Evaluation: A Compendium of Data for Emergy Computation*, Center for Environmental Policy, University of Florida, Gainesville, FL, 2000.

[ODU 00b] ODUM H., BROWN M., BRANDT–WILLIAMS S., "Introduction and global budget", in *Handbook of Emergy Evaluation*, Center for Environmental Policy, University of Florida, Gainesville, FL, 2000.

[OZG 07] OZGENER O., OZGENER L., "Exergy and reliability analysis of wind turbine systems: a case study", *Renewable and Sustainable Energy Reviews*, vol. 11, pp. 1811–1826, 2007.

[PAO 08] PAOLI C., VASSALLO P., FABIANO M., "Solar power: an approach to transformity evaluation", *Ecological Engineering*, vol. 34, pp. 191–206, 2008.

[PAU 14] PAUDEL S., SANTARELLI M., MARTIN V. *et al.*, "Wind ressource assessment in Europe using emergy", *Journal of Environmental Accounting and Management*, vol. 2, pp. 347–366, 2014.

[PED 92] PEDERSEN T., PETERSEN S., PAULSEN U. *et al.*, Recommendation for Wind Turbine Power Curve Measurements to be Used for Type Approval of Wind Turbines in Relation to Technical Requirements for Type Approval and Certification of Wind Turbines in Denmark, Danish Energy Agency, 1992.

[RAU 12] RAUGEI M., "A different take on the Emergy Baseline – or can there really be any such thing?", *Biennial Emergy Evaluation and Research Conference*, University of Florida, Gainesville, FL, 2012.

[REZ 14] REZA B., SADIQ R., HEWAGE K., "Emergy-based life cycle assessment (Em-LCA) of multi-unit and single-family residential buildings in Canada", *International Journal of Sustainable Built Environnement*, vol. 3, pp. 207–224, 2014.

[RIE 74] RIEKERT L., "The efficiency of energy utilization in chemical processes", *Chemical Engineering Science*, vol. 29, no. 7, pp. 1613–1620, 1974.

[ROM 99] ROMITELLI M., "Emergy analysis of the new Bolivia – Brazil gas pipeline", *Proceedings of the 1st Biennal Emergy Research Conference*, Center for Environmental Policy, University of Florida, Gainesville, FL, pp. 53–69, 1999.

[SAH 06] SAHIN A. D., DINCER I., ROSEN M. A., "Thermodynamic analysis of wind energy", *Energy Research*, vol. 30, pp. 553–566, 2006.

[SAL 76] SALONEN K., SARVALA J., HAKALA I., *et al.*, "The relation of energy and organic carbon in aquatic invertebrates", *Limnology and Oceanography*, vol. 21, no. 5, pp. 724–730, 1976.

[SCI 10] SCIUBBA E., "On the second-law inconsistency of emergy analysis", *Energy*, vol. 35, pp. 3696–3706, 2010.

[SEG 14] SEGHETTAA M., ØSTERGÅRD H., BASTIANONI S., "Energy analysis of using macroalgae from eutrophic waters as a bioethanol feedstock", *Ecological Modelling*, vol. 288, pp. 25–37, 2014.

[SHA 16] SHARMA M., SINGH O., "Exergy analysis of dual pressure HRSG for different dead state and varying steam generation states in gas/steam combined cycle power plant", *Applied Thermal Engineering*, vol. 93, pp. 614–622, 2016.

[SHE 10] SHEN L., WORRELL E., PATEL M., "Open-loop recycling: A LCA case study of PET bottle-to-fibre recycling", *Resources, Conservation and Recycling*, vol. 55, no. 1, pp. 34–52, 2010.

[SHU 10] SHUNPING J., BAOHUA M., SHUANG L. *et al.*, "Calculation and analysis of transportation energy consumption level in China", *Journal of Transportation Systems Engineering and Information Technology*, vol. 10, pp. 22–27, 2010.

[SZA 02] SZARGUT J., ZIEBIK A., STANEK W., "Depletion of the non-renewable natural exergy resources as a measure of the ecological cost", *Energy Conversion and Management*, vol. 43, pp. 1149–1163, 2002 .

[SZA 05] SZARGUT J., VALERO A., STANEK W. *et al.*, "Towards an international legal reference environment", *Proceedings of ECOS 2005*, Trondheim, Norway, pp. 409–420, 2005.

[TEN 88] TENNEBAUM S., Network Energy Expenditures for Subsystem Production, PhD thesis, University of Florida, Gainesville, FL, 1988.

[USG 15] USGS, *USGS Minerals Information*. USGS Science for a changing world, available at: http://minerals.usgs.gov/minerals/pubs/myb.html, 2015.

[VAL 02] VALERO A., RANZ L., BOTERO E., "Exergetic evaluation of natural mineral capital", *Proceedings of 15th International Conference on Efficiency, Costs, Optimization, Simulation and Environmental Impact of Energy Systems*, Berlin, Germany, pp. 54–61, 2002.

[VAL 09] VALERO A., BOTERO E. VALERO A., Global Exergy Accounting of Natural Resources, EOLLS Encyclopedia of life support systems, 2009.

[VAL 15] VALERO CAPILLA, A., VALERO DELGADO A., *Thanatia: The Destiny of the Earth's Mineral Resources*, World Scientific Publishing, 2015.

[WCE 87] WCED, *Our Common Future*, Oxford University Press, Oxford, 1987.

[WIL 12] WILFART A., CORSON M.-S., AUBIN, J., "La méthode EMERGY : principes et application en analyse environnementale des systèmes agricoles et de production animale", *INRA Productions Animales*, vol. 25, pp. 57–66, 2012.

[WU 15] WU X., WU F., TONG X. *et al.*, "Emergy and greenhouse gas assessment of a sustainable, integrated agricultural model (SIAM) for plant, animal and biogas production: Analysis of the ecological recycle of wastes", *Resources, Conservation and Recycling*, vol. 96, pp. 40–50, 2015.

[YAO 10] YAO M., WANG H., ZEANG Z. *et al.*, "Experimental study of n-butanol additive and multi-injection on HD diesel engine performance and emissions", *Fuel*, vol. 89, pp. 2191–2201, 2010.

[ZHA 10] ZHANG G., LONG W., "A key review on emergy analysis and assessment ofbiomass resources for a sustainable future", *Energy Policy*, vol. 38, pp. 2948–2955, 2010.

Index

Printed in the United States
By Bookmasters